THIS LIVING EARTH

How many grass plants does a single doe span as she
stands browsing at the edge of a field?

For how many insects and spiders is that space between
her four hooves an entire world for a day, or for a
lifetime? How many minute crawlers of the dark
know as their entire universe only the labyrinthine
root web of a single clump of grass?

Lie in a meadow, explore, discover, and you will find
that there are worlds within worlds.

THIS
LIVING EARTH

Words and Photographs by David Cavagnaro

AMERICAN WEST PUBLISHING COMPANY
PALO ALTO · CALIFORNIA

This book is dedicated to Maggie,
who shares the thrill of discovery
and knows the meaning of joy.

©Copyright 1972, American West Publishing Company. Printed in the United States of America. All rights reserved. Reproduction of the whole or any part of the contents without written permission from the publisher is prohibited.
Second Printing, July 1973.
Library of Congress Card Number 72-75555

ISBN 0-910118-25-6

CONTENTS

PROLOGUE

A few years ago my wife and I discovered a meadow. It was not unlike others we had walked through many times before.

I knew a little about meadows already because I had grown up on the fringe of town. As a child, I loved to explore. I hiked a lot and learned something about the out-of-doors. I began collecting insects when I was seven, raised caterpillars, and watched the activities of pond life in an aquarium I kept by my window. About meadows, I learned that they were busy, beautiful places and that I felt just plain good being there. This much I knew from my youth, when impressions meant a great deal.

Maggie knew this, too, but for the most part she learned it later. She grew up in the city, where most of the greenery was planted and even a grassy weed lot was uncommon. Something must have happened, though, when she was a child—maybe a weekend in the country, or the discovery of spiders or snails in her backyard. Most tendencies in one's life have small and strange beginnings.

Later, in college, it all came flooding back to her, the smell of damp grass covered with dew, the awareness of other creatures besides ourselves with lives to lead and problems to face. Biology classes, weekend field trips, a summer course in the mountains, and Maggie was wide-eyed with wonder and excitement.

This particular meadow was different only because we ourselves were finally receptive. It could have been any meadow anywhere, or a forest, a pond, or the seashore. The first walk we took was memorable because it was a beginning, the first in a long series of discoveries, as important to us as the first object an infant crawls to under his own power. The place could have been any place, but the time could not have been any other time.

THIS LIVING EARTH is the story of three years during which Maggie and I became acquainted with the grasslands of California. Grasslands are forgotten landscapes, as familiar as a backyard weed patch, a vacant lot, a meadow, or a grassy hill, yet as unknown to most of us as the wildest jungles. At one level, this book is a journey among the fascinating creatures that inhabit the grasslands and an investigation of the ecological relationships between them. On another it is an exploration of ourselves, for during these three years we became more aware than ever before that learning is an evolving process of discovery and that the web of knowledge, just as surely as the web of life, consists of many parts spun together in one immense, dynamic whole.

The first morning walk in our nameless meadow was the beginning of an exploration in the land of the small. What we experienced was a stretching of the senses and the spirit more than of the limbs, a rediscovery of a simple, primitive awareness, and we shall never forget the surprise, the thrill, and the joy of it.

AWAKENING

By crawling through the grass with all senses alert,
we found in ourselves a simple, primitive awareness
of the living earth

One morning in early spring Maggie and I awoke at dawn with an irresistible impulse to break through the rigid structure of our routine. Was it simply the need for change? Or was it an inner knowledge, or at least a suspicion, that there is more to life than we normally assume, just as there must be more in a seed than a dormant embryo, and more in a flower than our own perception of its beauty?

We were both tired of wintering. The birds were singing, and the air was cool and inviting. The long spell of hibernation had somehow been broken.

No thought was given to planning. I grabbed my camera, and we jumped into the car. We drove beyond the edge of our small town, past the gas stations, past the last cluster of apartments filled with commuters still asleep because it was Saturday, and on through a rolling green countryside where dairy cows and horses grazed and the fences were gray with lichens.

We drove until the land looked wild, until we had gained enough distance to feel free of our normal attachments. We found an old side road and stopped. It was a good beginning, better than we could have planned—a broad meadow with scattered oaks, a small creek lined with willows and alders, and a great forest of firs and bays rising up and over the hills beyond.

A modest hike was our intent, to stretch our winter-stiff limbs and see some country, and we set off at a fair pace. We never got past the meadow.

It was no reflection upon the forest or the stream or the willow thicket that the meadow arrested our attention that early Saturday morning. It could have been any of them, for every wild place is filled with inspiring sights and incredible things happening. The rest of that magnificent landscape remained unexplored because of a simple accident of time and place: we got to the meadow first.

The sun was just coming over the distant east ridge, filtering through a line of trees on the crest, then breaking free and rising as the earth turned into another day. For the planet itself, this was just another spin added to the billions upon billions that had gone before it. Its rocky matrix, poking through here and there like mountain peaks—even as a few rocks scattered about the edges of our newly found meadow—would receive the warm rays and expand a little; a few more molecules would be loosened, and a few more crystals would begin to crumble. When it was all over, when the spinning had left the sun behind, beyond the ocean horizon, and its rays were working on the bones of the Alps and the Himalayas,

Any wild place is filled with incredible things happening.

Spring is
a wet time
in the grassland.

and upon meadow stones of other continents, the earth would be just one day older. A flicker of time would have passed which, in the larger order of the universe, would mean almost nothing.

But for all the living things in the meadow, this instant of time was of the utmost importance. It was an awakening. The birds were singing, welcoming the arrival of the sun. Robins were exploring the wet grass, stopping, waiting expectantly for just the right moment to catch an earthworm off guard. The trees, the big, spreading oaks with a new blush of spring green haloed in the early rays of light, made no sound, but we knew enough about leaves to understand that something was happening in the chemistry of their cells which only happens in the light of the sun, and we knew that even the trees were waking up to a new day.

Through the remainder of the spring, Maggie and I revisited our nameless meadow many times. We went there most often in the early morning because we sense a special magic in those first hours after daybreak. It is a time of transition, pregnant with change, alive with an insistent stirring prompted by the sun.

The spring passed quickly. A few hours spent with the dawn made light work of the rest of the day, and time whizzed by. Something good was happening to our energy. We felt revived and very much alive. And no wonder; routine had little grip on us that spring. Every weekend morning visit was different and filled with little discoveries.

Some days were overcast, and the light was gray and somber. When it rained or drizzled, we sometimes went anyway and sogged through the wet grass, following the chirping of tree frogs or watching the robins catch worms. But when the sun rose warm and clear over the east ridge, the mornings were at their best, and the meadow was ablaze with gems of dew.

No matter what the weather, we discovered, spring is a wet time in the grasslands. We came home with wet feet after every walk and often were half soaked from crawling through the grass after some new ground-level tidbit. It seemed that shirts and pants were always drying by the heater, no less, either, when the weather report read "fair skies."

During our earlier visits, when the mornings were still brisk, we spent most of our time watching the mist rise and the dew slowly evaporate. It was a one-color world then, and the tiny variations in shades of green were more subtle than our eyes had learned to detect. We were not yet aware of the many kinds of leaves comprising that broad meadow carpet, nor did we realize the promise they held.

We weren't long in learning. One morning we found a few scattered buttercups spreading their waxy, golden petals toward the sun. A week later there were more, and by the time another week had passed the meadow was a shimmering, vibrant basin of yellow flowers.

The buttercups were not alone. Beneath the oaks and again along the wooded margin of the meadow, we found baby blue-eyes opening, unfolding their five petals and five stamens in perfect symmetry. They were fragile and easily bruised. On one visit after a heavy rain, we found them tattered and limp, beaten to the ground and soggy from heavy tree drip draining from the branches above. When we first discovered them, though, and at other times later, they were wet only from the light touch of dew, their delicate, pale blue petals studded with tiny drops of water. As we lay right next to the blossoms in the wet grass, we could see how an occasional sunray passing through the trees defined every vein and even the cells themselves, so thin were the petals.

We soon discovered that we were not the only creatures attracted to the colors of these first spring flowers. Small wild bees with coats of gray or brown hairs and flies of various kinds were busy finding the blossoms, too, and they darted quickly from one to another as the morning sun warmed the meadow. As the season advanced, multitudes of insects appeared in the meadow. Soon the California poppies were in bloom, replacing buttercup gold with brilliant orange, and new bees were on the wing in search of them. Crane flies blundered along in front of us as we stirred up the grass with our walking. Their long legs dangled, and although their wings beat furiously, they made slow progress and never flew far.

Each morning we saw something new. During one visit, when a strange yellow *Orthocarpus*—known for some mysterious reason as Johnny Tuck—was blooming profusely across the meadow, we found the first katydids of the season. They could not have been more than a week or two out of the egg, for they measured less than a quarter of an inch long. We saw several sitting on the puffy little blossoms of Johnny Tuck, basking in the sun, cleaning dewdrops from their appendages, wiping their compound eyes with their front legs as though either the night had been too long or the sunrise too bright.

Baby locusts appeared not long after the katydids—short, stubby little things, mostly head and legs, yet unmistakable miniatures of the winged grasshoppers they would in due course become. Butterflies, beetles, and wasps joined the bees and flies at the flowers. Caterpillars munched their way through a rapidly changing and varied meadow fodder, and it wasn't long before another generation of butterflies graced the fields.

One time we had the good fortune to find a remarkably beautiful sphinx moth, one of the smallest of our hawkmoth species. It was newly emerged; not even the most minute blemish could be seen on its wings. It rested in the grass, its green and gray fore wings held back like those of a jet plane. It seemed the epitome of streamlining; no wonder hawkmoths are such fast fliers! I reached out and touched the moth gently. It raised its front wings in protest and revealed a pair of orange hind wings which rivaled the nearby poppies in brilliance. Maggie and I were both startled by this sudden display, and we wondered if a hungry bird searching for a meal might not be startled, too. By the time we had collected our wits, the moth was gone, and the meaning of the orange wings became clear.

By the end of May changes in our meadow were piling up against each other, accumulating faster than we could keep track of them. We sensed a tension, an urgency about time, that hadn't been present in the spring. The feeling was purely our own for we knew that the plants and insects which had been our companions through the spring could not, in the human sense, be aware of the passage of time.

Yet all the organisms of the meadow were responding in their own ways to physical changes that impinged upon them as the days

There is a sense of urgency about summer.

15

passed. They were far more responsive than we were to the fact that the sun rose each morning a fraction farther north along the east ridge and crested higher in the sky at noon. They responded to the gradual lengthening of the days and warmer temperatures, changes we scarcely noticed because of the insulation that enveloped us during our waking and sleeping hours.

Every living thing in the meadow knew that the rains had ceased, and they knew it without benefit of the power of thought, without the aid of weather reports. In myriad, subtle ways their complex chemistry was responding to the stimuli of temperature, moisture, and light. They knew, far more directly than we, that summer was coming.

Every living thing in the meadow knows when the rains have ceased.

I suppose that Maggie and I must have commented occasionally about sunny skies, but we were aware of summer in other ways peculiar to our culture. School would soon be out, and our teaching programs would end; we counted the days by watching the calendar. The water bill was slightly larger, the heating bill slightly smaller. Friends were busy unpacking musty tents and sleeping bags in need of an airing. Everywhere people were preparing for vacation.

But in our meadow we watched summer advance by discovering the effects it had already produced. We learned about summer by seeing its results. A few things we noticed directly. The mornings were already warm by the time we began our walks, even though, by the clock, we arrived much earlier than before. There was scarcely any dew, and we came home dry from our belly-crawling excursions. My camera no longer fogged up around the viewfinder, nor were my fingers too numb to make the settings easily.

The flowers of early spring were gone, and those that followed them were fading. Owl's clover and lupine had bloomed to the ends of their stalks; only a faint blush of pink and blue remained where great, brilliant swaths had once prevailed. Ephemeral ponds and boggy places in fields near the meadow were drying, and the creamy patches of meadow-foam that had lined them were well past their prime. Even the sunflower heads were fading on the mule-ears clumps growing on rocky knolls, and the displays of poppies with them were reduced to scattered dots of orange.

The grass, once a carpet jeweled with dew, had risen into the breeze. The stalks pulsed in broad, sweeping waves; they opened their tiny flowers and cast their pollen into the wind. Had either Maggie or I been susceptible to hay fever, our morning walks would have been far less pleasant than they were.

Signs of summer forged across the land. The meadow changed so rapidly that our weekend trips were like reading only one page from each chapter of a book. The production of seeds was one event that was well under way when we finally read of it among the stalks and drying blades of the meadow.

It must have been happening long before we became consciously aware of it. In spite of our practice all through the spring of looking closely, lying in the grass for a better view of this diminutive world, we had somehow missed the seeds or taken them for granted. Intellectually, yes, I suppose we knew they were there, for if someone had asked us if the time had come for the setting of seeds, we would have answered correctly. We simply hadn't seen

16

The first morning walk Maggie and I took in our
nameless meadow was the beginning of a new
exploration. What we experienced was a stretching of the senses
and the spirit more than of the limbs—a rediscovery
of a simple, primitive awareness—and the joy of it will
never be forgotten.

We visited the meadow often through the spring.
Dawn was the most exciting time, for every
spider web and blade of grass was jeweled with dew.

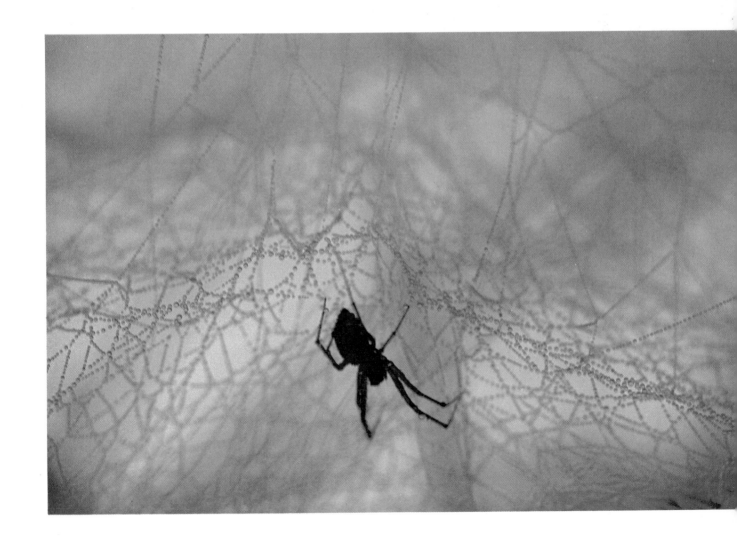

The green carpet of spring was soon broken.
In the meadow buttercups appeared, and
beneath the oaks we found baby blue-eyes opening,
unfolding their five petals and five stamens
in perfect symmetry.

Week by week, as spring advanced, the ranks of flowers changed. Soon California poppies were blooming, replacing buttercup gold with brilliant orange.

We began to notice insects appearing in the meadow—a newly emerged sphinx moth,
the first infant katydids, many kinds of butterflies, and numerous wild bees visiting the flowers.

them with a knowing eye, even though we must have been looking at the first of them ever since the buttercups and baby blue-eyes had stopped blossoming many weeks before.

When we finally became aware of seeds in the meadow, we realized that we had been victims of preoccupation. As long as the flowers bloomed, wave after wave of them, and the insects were busy after their pollen and nectar, our attention had been focused on them. The brightest things in the meadow, the centers of action, had occupied us almost completely. They had blinded us from seeing the subtle happenings of the season. At first we felt rather angry with our attraction only to the spectacular. Now, in retrospect, we realize that we were simply engaged in an essential phase of the learning process; at the time we could only wonder what else we had missed in the meadow as spring yielded to summer.

Most of the color was gone from the grassy hills and fields when we made our first seed discoveries. Most of the flowers had performed their function. The deep-rooted California poppies were putting forth a second round of blossoms, paler and less showy than the first, and on the steeper slopes and road banks delicate pink *Godetias* were opening, signaling the change of season even in their common name, farewell-to-spring. But all told, the brilliance had vanished. The green, too, was being replaced by a faded brown, which crept from the dry, rocky slopes down across the valleys through the tall grasses of our meadow.

We traveled more then, stopping at other places before and beyond the meadow. We journeyed westward to the shore of the Pacific and eastward through range after range of coastal hills. We crossed the great Central Valley of California and explored the lower foothills of the Sierra Nevada. We traveled through fields and hills covered with grass, many kinds of grass, many textures, many shades and degrees of summering showing in the color of the stalks, and only where forest or brush grew, where streams nourished willows and cottonwoods, or where man had built or plowed were there interruptions.

These were the grasslands—open prairies, wild and windy, rolling gently away toward clean horizons broken only by an occasional windmill, a farmyard clump of trees, or a power line; dry hill prairies of the inner coast range, steep and folded into ridge upon ridge of brown grass running down to the edge of the rich river alluvium where the plows of the Central Valley farmers had stopped; oak grasslands of the coastal hills and the lowest slopes of the Sierra, where the great trees grow together in groves or individually, old giants twisted and sometimes broken, their limbs drooping almost to the ground; grasslands of ancient sea benches, facing straight into the winds and fogs of the Pacific, winds that defy even the pines and keep them back against the steeper slopes.

These were the grasslands of California, and we wondered about all the others—the windswept prairies of the Great Plains, where the searing blasts of summer follow the blizzards of winter; the grasslands of the eastern states, green from summer rains and dormant when the weather turns cold; the steppes of Eurasia; the savannas of the tropics; and mountain meadows on every continent. With or without trees, grasslands dominate the world's vegetation. As we drove from place to place, we talked a lot about grasses, and our appreciation of them grew.

With or without trees, grasslands dominate the world's vegetation.

26

It grew more from seeing them closely, though, than it did from any armchair (or car-seat) intellectualizing. No textbook stories about the wonders of grasses could have replaced what we learned during that first summer of exploration. There is no substitute for smelling dry grass; for watching its stalks in uncountable numbers rippling in the wind as it blows across the face of a hill; or for seeing them singly, studying their graceful forms, feeling their varied textures, or pulling their sharp seeds from the weave of socks.

The seeds were forming fast, and many had already been shed by the time we extended our travels beyond our meadow. The mornings were warm even before sunrise, and the sun itself was blistering. As time passed, it became harder and harder for us to imagine that all the other meadows and hills we explored had recently been green, too, and covered with dew in the mornings, as our meadow had. The tule fogs of spring were only wispy memories. We ourselves were transformed as the grasslands were transformed; we were lazy and moved slowly. The heat was oppressive. By midmorning we were ready to sleep in the shade of a spreading oak, and at night, when the sun had gone but not the heat, we sat up in camp for hours listening to the crickets calling.

But in the mornings, early, we walked miles through the dry grass, and then the only sound was the snapping of lupine pods firing their round, black seeds. We were aware of seeds finally, and we found many kinds. There were burrs and barbs that stuck annoyingly in socks and pants legs. And there were pods, too, besides those of the lupines, pods of many shapes and sizes, each filled with a precious load of seeds. The flowers and grasses, except for those with deep, perennial roots, were crisp and dead. In their dormancy, the seeds and roots were the only link with the future, the only potential for next year's growth. The grasslands were sleeping, and the summer sun blazed day after day.

The plants were not alone in their summer sleep. Where were the flies and bees that had droned over our meadow? We returned to that place of our initiation and listened, but we heard only long intervals of silence broken now and then by the loud buzz of some insect in a tremendous hurry. Where were the caterpillars we had seen chewing on leaf and blade, or even the butterflies and moths they had become? Where had the wolf spiders gone, and all the spinning spiders whose webs we had seen laden with dew?

The air in our meadow was dry and still. Hardly anything moved where thousands of springtime residents had once flitted and crawled. Except for the fragile, slow-flying ringlets and the darting buckeyes, butterflies were scarce. The only insects in abundance were the grasshoppers. They had recently reached adulthood, and as we walked through the grass, we flushed dozens and sent them flying ahead of us in the pattern of a fan. The clicking of their wings sounded flat in the stillness and blended with the sound of dry grass crackling under our shoes.

Many of the smaller organisms of the grasslands had apparently finished their activity for the season. As eggs or larvae, pupae or adults, they must have hidden themselves in some protected place and lapsed into dormancy. Deep in earth cracks, under bark flakes, in the cool

In their dormancy, seeds and roots are the only link with the future.

27

As signs of summer forged across the land, we journeyed beyond our meadow, from the shore of the Pacific to the foothills of the Sierra Nevada. Everywhere the grasslands of California were browning beneath the blazing sun.

hollows of rotting oak limbs, they would carry on their sleepy estivation, waiting out the long summer as hibernating animals wait for the passing of winter.

Not all grassland animals were dormant, however. On trips farther from home we camped for two or three days. The nights were balmy, sometimes too hot for sleeping, and we explored the fields by flashlight. Once in a while a moth would ghost across the beam of light, and we could see the eyes of wolf spiders glinting by reflection. We tried to find the crickets that made constant music in the night, but they knew we were there long before we could spot them, and stopped their chirping.

These were summertime creatures. They simply avoided the heat of the day by being active at night. They were not adapted for withstanding the desert-like heat and dryness of the grasslands as the locusts were. Instead, they evaded the worst of it, and at night, as we discovered, they were on the prowl.

We made another interesting discovery one evening late in summer. It was a discovery more about ourselves than about the creatures whose lives we were following through the seasons. A jackrabbit leapt out of the grass in front of us and jarred loose the realization that we had been fascinated with the small and minute and had missed the larger creatures of the grasslands. Once again we had been blinded, as we had by the flowers of spring. The hare bounded away into the dusk. The sudden explosion from nearly beneath our feet had startled us; it had awakened us, too, and that night in camp we sat pensively for a long time considering what other animals we might have missed during our many little treks through the grass.

Answers came faster than we expected. We had been quiet for a while when, almost simultaneously, we caught sight of movement along a small gully on our right. It was a skunk stalking boldly along the defile, possibly hunting mice or insects. And before we went to sleep we saw the faint silhouette of an owl gliding low overhead, passing silently through the twilight.

Through the spring and summer we had crawled about with our noses to the ground. Our newly found perspective had greatly broadened our horizons, but had limited them too. There is something confining, we learned, in a particular frame of reference, no matter how inspiring and mind-expanding it happens to be. I was reminded of an experience I had had years before in Australia while traveling in search of insects. I had turned over a log and found several species I had not seen before: three or four kinds of beetles, a few termites, a colony of ants. I was so absorbed with these discoveries that while collecting them I failed completely to notice a very large lizard resting directly beside my elbow! So it is with vision; seeing is closely allied with the center of interest and the concentration of the moment.

We made a conscious decision that "night of the jackrabbit," as we have since called it, to spend some of our walks searching for things beyond the grass-roots focus, and we were not entirely disappointed.

We saw red-tailed hawks circling the updrafts above the hills, and we watched kestrels hovering over the fields in search of mice or grasshoppers. When the larger hawks invaded their territory, the sparrow hawks would dive upon them in mid-air and chase them away. We found colonies of ground squirrels and

Nocturnal creatures adapt to summer by evading the worst of it.

enjoyed their whistled warnings that danger was near. In the evenings we saw owls many times, and gray foxes, more feline than canine, prowling shyly along the meadow margins.

One night, as we lay camped in the foothills of the Sierra under a waxing moon, we heard coyotes screaming beyond a far ridge. We were saddened by this reminder of the glorious animals of hill and prairie that have fallen before the onslaughts of man. Coyote, antelope, elk, bobcats, mountain lions, even bear, were common before the coming of the white man. The grasslands are empty of them now. They linger only as ghostly shadows of the imagination, as fanciful legends recalled around the campfire, or as rare, eerie cries in the night from some far-off crag beyond range or interest of the hunter.

In this instance our search for the larger creatures had been a disappointment. For many who had gone before us, conquering the wilderness had been a source of great pride. We were now going more softly across the land than they; our trophies were not measured in skins or heads. By their conquest life in the grasslands had been gravely diminished. For us, the taste of their pride was bitter, and the cries of the coyote seemed hollow and lonely.

By the end of summer most of the creeks had dried up completely. Tender leaves and water were both scarce, and herds of deer were on the move. Evening after evening we watched them grazing in the dry fields, searching for food beyond the cover of woods and scrub. We were puzzled that the deer were drawn into the open and that they seemed to be finding something to eat where we could see only withered stalks. We made a careful search of the fields, and though we saw nothing that showed signs of being eaten by the deer, we did find numerous young leafy plants growing as in spring.

We knew what they were the moment we walked through a patch of them. A pungent aroma greeted us in the warm air, and our shoes and pants legs were soon sticky. We had found the tarweeds, though only a faint tinge of green appeared through the brown grass and no flowers yet revealed their presence from a distance.

What amazing plants they are! When everything else in the grasslands is brown and dormant, when the soil is dry and deeply cracked and scorched by the summer sun, the tarweeds are green and growing. Their leaves are hairy and oily to reduce the loss of moisture, and their roots grow deep. Somehow they find enough water in the parched fields months after other plants have given up the search.

Although the tarweeds are practically the only green plants in the summer meadows, the deer and rabbits rarely touch them. It must be that the hairs and volatile oils which help protect them from desiccation also render them unpalatable, and they are left alone to reach maturity. What the deer were actually finding in those dry fields remains a mystery. Perhaps they were eating the dead leaves and stalks after all, to tide them over until the first rains brought forth a new crop of tender greens.

A few weeks later the tarweeds began to flower, and soon the fields were golden. They had the grasslands to themselves, and special kinds of bees appeared on schedule to pollinate them. The tarweeds were the final fling of

By man's conquest, life in the grassland has been gravely diminished.

The stiff seed barbs of the grasses, the
graceful pod of the mariposa lily, the sound of
lupine pods snapping were signs of summer sleep
that crept across the land. The grasslands
were dormant; the air was dry and still.

summer. September had arrived, and with it the hottest days of the year. The fields overpowered us with tarweed smells, and we soaked them in as fine perfumes. We soaked in the heat, too, for we knew that soon the rains would return.

We visited our original meadow in late September after a long absence. Thistles were shedding their seeds; the air was filled with their soft plumes, and the sticky tarweeds were white with thistledown. It was a magic moment. We sensed an urgency about time, as we had in the spring, only now the inhabitants of the grasslands seemed tense with waiting. In most quarters, there was still no flurry of activity, no great rush to get things done. Hardly anything moved in the meadow except a flock of birds hunting thistle seeds, yellow jackets droning in the grass, and the seed plumes of the thistles drifting in a gentle breeze.

Each morning the sun rose farther south along the east ridge. The oaks were laden with acorns; squirrels and jays were busy gathering winter stores; the rat-tat-tat of acorn woodpeckers echoed across the valley. Vacationers came home from their last spree, and the kids went back to school. Hayfields were cut, and the bales neatly stacked, pumpkins ripened in country gardens, and dry stalks of corn rattled in the hot wind.

Indian summer, we always called it. A change was coming, just as surely as it comes with the warmth of spring to the grasslands of the eastern states. Some creatures were busy—those that had acorns or hay bales to gather, families to feed, or lessons to learn in school or grassy fields. But most simply waited, especially those sleeping in the grasslands, for the one thing that could waken them: the coming of the rain.

We knew from long experience that the first rain would probably fall in October. The breaking up of the summertime pressure system, the atmospheric changes that finally allow Pacific storms once again to move against the western edge of the continent, and more specifically, the coming of the rain itself would drastically alter the look and feel of the land. This much we anticipated, and we eagerly awaited the falling of the first drops.

But weather is fickle and full of surprises. This particular year a cold front moved in over the land before the first rain; the nights were suddenly brisk, and heavy tule fogs settled in every little valley and hollow. It was a subtle thing. We would not have noticed that the chill came before the rain rather than after it, had it happened the year before. At least we would not have experienced its fantastic results. But over the year we had changed. We were aware of events that would have escaped us before, and on that first cool morning we went to our meadow as though we had been beckoned. No voice called us but the silent voice of wonder, the keen pulse of curiosity that had been growing within us through three seasons.

The fog was dense and wet; the chill seemed intense after the heat of Indian summer, and we weren't dressed for it. We shivered as we drove past the edge of town into the country beyond. When we arrived at the meadow, the sun was just lifting above the east ridge. The low blanket of fog began to part, and the warm rays streamed through. What lay before us was the most beautiful scene we had ever witnessed; or maybe we had seen such things before but had not been ready, like a child who hears a word many times before he understands its meaning.

The tarweeds, the whole meadow, wore a

mantle of thistledown, but not as we had seen it before. During the night the mist had settled and condensed, and every tiny plume was jeweled with droplets of moisture. Strung between the tarweeds and stalks of grass were hundreds upon hundreds of spider webs—mazes of strands, suspended sheets, large orbs—every one heavily laden with dew. As the sun rose over this magic creation of the night and scattered the mist, the meadow blazed as though it were illuminated by billions of microscopic lights. Where we had seen only dry weeds and thistle plumes before, we now saw a small world of unsurpassed beauty, fashioned in a single night by the simple interaction of temperature and moisture, the productions of living things, and the happenstance of timing.

In the normal course of events, the rain comes first and cools the land. After the autumn storm has passed and the nights are clear, the spiders rebuild their webs, and the meadows glisten with dew in the light of sunrise. But the parachutes of the thistles cannot be rebuilt. They do not appear again in the fields until another year has passed. These fragile plumes are made for the wind; the first rain of autumn destroys them, rips them from the tarweeds, pounds them into the grass. When the dew finally comes to the autumn fields, the thistledown is gone, even the grasses themselves are broken and matted, and only their tougher stems remain upright to bear the drapery of the industrious spiders. The webs alone gather the tiny droplets and sparkle in the morning sun.

We walked silently into the meadow, stepping carefully to avoid the best displays. Everything was still intact; no rain had damaged the delicate leftovers of summer. We discovered fine, soft grasses with beads of water on every spine and hair. In several places we found orb webs that had not been torn by insects and rebuilt during the night. A large banded garden spider rested in the center of each, unaware, or unconcerned, that thistle plumes had lodged in its net, trapped from the breezes of the previous afternoon. I suppose that a bit of fuzz stuck in a spider web is of little consequence among the larger affairs of the world, but to Maggie and me that morning, discovering these two finely wrought creations of nature juxtaposed and studded with dew seemed like an experience of a lifetime.

The morning had an unmistakable feeling of autumn. The dampness, the fog, the soft touch of mist rising against arms and cheeks suntanned from many summer walks, and boots soggy for the first time since spring—these were part of it. But more than anything, the feeling of fall was born in the air as the sweet smell of moisture on dead grass. We breathed deeply and took in the perfumes of morning. There is no describing those aromas, except perhaps as a tremendous relief from the dust and dryness of summer. The spell had been broken, and we were glad for the change.

We continued our exploration of the meadow, revisiting all the spots that had become so familiar to us over the year. Apart from the dew, there was little visual evidence that summer was edging into fall. There were no great displays of colored leaves, none of the traditional signs of the season. The maples and willows along the stream were still green, and the grass was a fairly even, somber brown.

As we walked, though, we found hints of

Indian summer offers hints of impending change.

35

During the hottest part of summer, in what
seemed an incongruous display of growth, the
tarweeds burst into full flower, a new set of
bees appeared to pollinate them, and black
blister beetles came to chew their petals.

impending change. The sheep sorrel was bronzing, and a few yellow leaves had fallen from the willows. Along the edge of the meadow bordering the stream we discovered some leaves fallen from the vines of wild grape that tangled through the trees. The leaves were scarlet in the backlight of morning, and dead grasses cast geometric shadows against the mosaic pattern of their veins; they were dying in splendor. They blazed in the grass, and we read them as symbols that death was somehow right, that a full season's function had been fulfilled and the vine from which they came was ready for a rest.

By the time two hours had passed, the morning was transformed. The sun was high and warm. Thistle plumes had dried; spider webs had returned nearly to the realm of the invisible. The sparkle was gone, and the dry grasses rustled in a gentle breeze. The meadow looked much as it had during the weeks before. The sun had nudged summer back into power, and autumn seemed to retreat into those few brilliant leaves lost somewhere in an immense field of brown grass and thistledown.

For two priceless hours Maggie and I had been transported ahead of schedule into another season. People dream about time machines, but our small journey into time had been real. It was a journey inside ourselves, where traveling leaves its most permanent impressions.

The world we had seen that October morning was a world no human being could create, but it was one every human being could enjoy. There was something in the meadow for all the senses, landmarks for as many personal journeys as there are people. They would need only to have their senses turned outward and their travels turned inward. Yet we were alone in our meadow, alone with our discoveries. Not another soul was out that morning to see what we were seeing, to experience perhaps the one moment in a lifetime when the thistledown was covered with dew.

Maybe this is a small matter, and who were we to judge the importance of anyone else's experiences? And yet we had a feeling, stronger even than it had been that spring morning when we began our journey, that there was something in this way of traveling, something in the process, that went beyond a field full of damp seed plumes, and was larger in importance than any discovery we had made.

As we left the meadow, the squirrels were busier than usual gathering acorns. Their chatter blended with the hammering of acorn woodpeckers chipping new holes in their storage trees. They had sensed autumn in the air, too, but they recognized it instinctively and reacted in predictable ways. Our reactions were quite different. They emanated from those uniquely human qualities that enable us alone, among all the animal forms of this earth, to project from the physical experience into the abstract.

We weren't sure what it was, but we knew that the importance was not in what we saw but in how we felt and what we had learned. It was the process of growth, not of body but of mind, that held the key.

The final change of season came only a few days after our thistledown discovery. A storm moved in from the Pacific during the night. We were awakened first by the wind and then by raindrops falling. The sound of them hitting the roof—a sound that had become a symbol of imprisonment and gloom by the end of the previous winter—caused only excitement and exhilaration now. Their gentle but persistent patter was welcome music. The land was parched, but the rain had come.

At the first light of dawn we were on the road to the meadow. The clouds were parting, and the air was still. Rain had fallen all night, and the spiders had not yet rebuilt their webs. The dried grass had been softened, and some of it lay crumpled in tangled mats. The last tarweed blossoms hung limp and soggy, and the thistledown had vanished beneath the grass. The crowning blow had been delivered against summer. With the coming of the rain only a one-way path remained, and it led straight toward winter.

But better than rain music in the night, better even than knowing that the long, dusty summer had ended, was the freshness of the cool morning air after that first rain. I've yet to learn whether the aromas that rise from the earth then are actually different from those that rise after all the following rains, or whether they simply seem so because our sense of smell has lost touch with them during their long absence.

As I think back to that morning and remember the smells of damp earth and wet stalks of grass rising with the mist at sunrise, I remember another experience I had several years before on the other side of the globe. I was traveling from India to Malaya on an old freighter and had been many days at sea. The open ocean has few odors, and those aboard ship were mostly of fresh paint and fuel oil. We came up to the island of Penang just after dark, and filled with the emotion of the moment, I wrote the following lines in my journal:

"On one horizon the lights of Penang sparkle on the flanks of the jungle island; far off to the west lightning flashes among distant thunderclouds. The full sweep of equatorial stars shine in a moonless sky. For the first time I have experienced the beautiful aromas of the jungle. Drifting out across the water, the smell of jasmines and spices fills the air; the sweet mustiness of damp leaves rolls and swells over the bay."

Never on land did I experience such fragrance from the jungle, especially when I was exploring in the midst of it. The first approach to a tropical shore from sea was a rare moment in my life, and it was made all the more vivid by the absence of earthy aromas on the voyage. In a sense, summer in the grasslands of California is very much a voyage between two distant shores, the last shower of spring and the first rain of autumn. Perhaps the fragrance of that autumn moment is only more potent because it is strange to us, and a little time is needed each year to become reacquainted.

As the sun rose higher over the meadow, clouds of insects lifted into the air in great smoky puffs. They drifted slowly upward on fragile wings glinting in the sun. In the squint of an eye they could have been giant snowflakes falling up instead of down, or another kind of seed plume parachuting in an unfelt breeze. All fantasies aside, however, we knew them to be insects, and we approached more closely to see what kind they were.

We found that they emerged from tiny holes in the ground, then crawled up the grass stems one after another in rather jumbled disarray. When they could climb no higher, they hesitated, balancing precariously until nudged into the air by the next in line.

Only man can project from a physical experience into the abstract.

39

These were winged termites, both kings and queens; we were witnessing the launching of their nuptial flights. It is a great event for the termite colonies, triggered only once each year by the first rain. We examined their burrow openings and watched in amazement as they streamed forth, scrambling two and three at a time through the narrow holes. They were surrounded by pale, wingless soldiers and a few workers, who had come out of their subterranean tunnels especially for the occasion. The mating flight is a perilous journey in a world alien to these tender creatures of the underground. The soldiers stood by to protect them.

We weren't long in discovering some of the reasons. The ants had found one emergence point and were busy dragging away some of the helpless kings and queens despite insistent protest of the soldiers. There was much activity with many losses on both sides of the struggle,

but the termites persisted by sheer numbers, and the air was filled with them.

Once aloft, however, those that escaped the ants met still another danger, for the flycatchers had also discovered their presence in the meadow. They darted back and forth through the swarms, feasting with great relish. But still the termites flew, and later in the morning we began to find them again on the ground, paired now and with their wings chewed off and discarded, the males following closely behind the females as they searched hurriedly for a suitable object to burrow under and begin a new colony.

It was an action-packed morning, filled with something new for all our senses. We were learning more about the seasons and about this country meadow without a name. We went home filled to the brim, as though we had been gone for months on a very long journey.

Summer
is a voyage
between
two distant shores.

By September, the monarchs were moving
from inland breeding grounds toward
the coast, where they would spend the winter.
Many would return to the milkweed fields
the following spring, but not all.

An early ground fog broke the spell of summer.
Delicate thistledown and grass tufts would
normally be destroyed by the first rain, but
instead we found them jeweled by the light touch
of fog. Soon, fallen leaves of wild grape
provided the first bits of autumn color
in the meadows.

The first October rain renewed the hills and meadows. The next morning, termites emerged for their annual mating flight, and within a week new grass was sprouting.

Once rains had cooled the grasslands, frost and tule fog welcomed the dawn. When spreading valley oaks greeted the sun with naked branches, we knew that winter had arrived.

49

There is no feeling quite like lying in the brittle grass
on especially cold mornings when the ground is frozen. Studying delicate
frost patterns and ice extruded from the soil like miniature arctic
"frost heaves" is a chilly but exciting adventure in the land of the small.

The last frosts of winter fringed the edge of spring;
the seasons had run full cycle. For a year
we had been fingering through pieces of knowledge
as children do when they spread the parts of
their first jigsaw puzzle before them,
not quite aware that in the sum of the parts
lies a picture yet to be discovered.

*When
the first rain
finally comes,
seeds stir
in their sleep.*

Autumn moved quickly for us. We traveled little, visiting mostly close to home. Rain fell frequently, and within a week of the first storm, the fields and hills blushed green with sprouting grass. We crawled low over the ground browsing through the tender blades. They were all there, all the seeds of summer germinating now in the moist mat of last season's discards. Not just the grass, but lupines, buttercups, poppies, thistles, and many other plants we knew from our summer walks were sprouting, too.

Perhaps the strangest thing about California is that in the grasslands the first week of autumn is, in effect, the first week of spring. The seeds stir in their sleep as the cool water penetrates. They soak it up and, almost overnight it sometimes seems, the countryside turns green. But unlike spring in most areas, this autumn spring of California is launched straight into the teeth of winter. No sooner are the tender shoots up and growing than the nights turn cold and the fields are mantled in frost. Every delicate leaf and blade bows under the weight of it, crisp and brittle with the cold. Yet they survive, for this is their land and they have evolved to meet its demands.

For a brief few weeks the maples, ash, and willows blazed golden along the stream courses; poison oak bushes and vines of wild grape turned scarlet and dropped their leaves. Huge flocks of cedar waxwings stopped in the trees on their way south for the winter, and band-tailed pigeons gorged themselves on madrone berries. Warm rains came out of the Pacific and passed on toward the Sierra Nevada, and in their absence, when the nights were clear, the frost worked its special magic upon the fields.

We spent many frosty mornings shivering in the grass, lying belly down on the frozen earth, photographing some delicate creation of the night or simply watching the tiny crystals melt as sunbeams probed the shadows. We found spider webs, and even fragile mushrooms, studded with frost, and we hunted for these early morning gems until our fingers and toes were numb and our ears roared with the sound of chattering teeth.

By January, cold storms out of the north replaced the warmer rains of fall. The creeks ran full, spilled their banks, and flooded marshes and low places in our meadow. Fierce winds blasted through the valleys, the last of the old grass was leveled, and some of the evergreen oaks lost their grip in the soggy fields. They seldom fell entirely; we usually found these old-timers of the meadows braced by an elbow against the storm, leaning half over, resting on a gnarled limb.

The magnificent valley oaks fared much better. Having lost their leaves in the fall, they offered less resistance to the wind. They stood naked against the billowing clouds, and beneath them lay a few old branches, victims of this natural storm-born winter pruning.

Between storms the frost would return, though there were many nights when a thick blanket of tule fog insulated the valleys and kept the temperature above freezing. The foggy mornings were incredible times to be out in the meadows. Spider web displays were always at their best then, and when the sun would rise through mist-veiled oak branches and the ephemeral pink glow of morning was soft and diffused, we could not imagine a better moment to be alive.

There seemed no end to the discoveries. Sometimes they were simply the same old

54

friends in different guises; at other times they took the form of some plant or animal, some pattern or texture, even a feeling, which we had not experienced before. However minor or major the discovery, it always came as a thrill and a surprise. After nearly a year of exploring the grass-roots landscape, there were still new things to find among the familiar. Learning this was an important discovery in itself.

Out of all that happened that winter, however, one experience stands out boldly from the rest. It was a Saturday morning. Torrential rain had fallen off and on for a week, but the clouds had dissipated just in time for a field trip I had planned with one of my classes. The hike was to begin in one of the meadows Maggie and I had visited many times, and by midmorning the group had begun to assemble.

When I arrived, I saw that those who were there before me were all staring intently in the same direction, some with binoculars. My eyes followed their gaze. I could swear that my heart skipped a beat when I saw the first flash of apricot fur. It was a weasel, busily exploring gopher burrows in the open meadow scarcely fifty feet from where we were parked.

The weasel was wary but made no sign of escaping. He would pop suddenly out of a hole, his beady eyes flashing in the morning sun, his round ears scanning toward us, and then just as suddenly he would disappear. These antics were still going on by the time everyone had arrived. His confidence encouraged us, and we crept slowly closer. For half an hour we watched each other. The weasel allowed us to approach within ten feet of his chosen retreat—and I didn't have my camera!

When it was all over and several days had passed, I pondered the immense feeling of disappointment I had experienced that morning. I had no record of that fabulous coat of orange fur, that rare moment when a secretive wild creature allowed our human intrusion into his private world. And yet the record was there, vividly impressed upon the nerve channels of my mind. The camera wasn't essential after all. It had become almost a crutch to help me see. The feeling was the important thing, that which came from the experience itself and could never be captured on film. What better tool does one have than his own mind, I thought, with all its senses fully tuned? Once again the meadow had spoken.

Imperceptibly, as the weeks passed, the grass blades grew taller in the meadows. The month of March brought the last blasts of winter against the western edge of the continent, and one morning we found the first baby blue-eyes open and ready for the business of spring. Soon the buttercups were flowering, fringed at dawn by the last frost of the season.

What a different winter it had been from the one before! No prison had caged us; our limbs were not stiff, and our minds were not dull. Something had happened to us. We had discovered a new landscape, a world of grasses; we had followed the grass through four seasons, and we had learned something about the life it harbors. We had learned to look closely and take nothing for granted because knowledge comes in strange and unexpected packages.

We had discovered more about this world now, and we knew more about ourselves. It was a good beginning.

Knowledge comes in strange and unexpected packages.

The meadow contains all the questions that man
has asked and many that he has not learned how
to ask, for the meadow with all its parts is
merely a tiny piece of the larger whole, a puzzle
within a puzzle.

DISCOVERING

Behind the hooting of an owl, the smell of damp earth, behind even the buzz of a mosquito lay the thrill and excitement of discovery

There are patterns in the grass, green blades crossing geometrically, spider webs covered with dew. There are rainbows in a single bead of moisture when the sun spills over the ridge, just as surely as there are rainbows of a billion drops in the sky after a spring shower.

There are worlds within worlds. If the range of the mountain lion contains hundreds of acres of woods and grassland, how many meadows does he see in each? How many grass plants does a single doe span as she stands browsing at the edge of a field? For how many insects and spiders is that space between her four hooves an entire world for a day, or for a lifetime? How many creatures creep blindly through the soil beneath my body as I lie in the grass, one organism, a single member of a single species staring at the woven catchlines of a solitary spider? How many minute crawlers of the dark know as their entire universe only the labyrinthine root web of a single clump of grass?

A year before, I might have had a vague feeling for such matters, born of gradually accumulating experiences and a sprinkling of knowledge gained in school. But the questions would not have been there, not close to the surface, not actually spoken in the verbal silence of the mind as they were now, not presented with the conviction of wanting to know. The love was there but not the commitment, and all the books in the Library of Congress could not have provided it.

A year of discoveries in the grasslands could, however, and it had. I stayed by the spider web for a long time, until the spring sun felt warm on my shoulders, and I thought back over the many trips Maggie and I had taken through the grass-tangled landscapes. I felt that we had seen much but missed even more. The addiction held us more firmly now; we were more excited over this spring than we had been over the last. But until now we hadn't been asking questions. Not many at least, nor with the earnestness which comes from seeing a patch of grass as a screen behind which hides the universe.

We were primed for the coming year. We had been fingering through the pieces, learning their form and color as children do when they spread the parts of their first jigsaw puzzle before them, not quite aware that in the sum of the pieces lies a picture to be discovered.

Nature is better, though, than any jigsaw puzzle. It is infinitely more complex, and there is no picture on the box to reveal the secret. The puzzle that nature offers cannot be illustrated; it goes beyond the scene itself—the trees and clouds and rocks and grasses that, together, form a landscape. It transcends the graphic. The

A patch of grass is a screen behind which hides the universe.

puzzle of nature is dynamic, multidimensional; it is a process, a flowing together of matter and energy, a series of interlocking systems both visible and invisible. The puzzle exists already assembled, but there is no way to see it, not all of it, with the eyes alone.

This was just a way of expressing in words, as I lay in the meadow, the feeling that beyond the grass there was another realm, a great river flowing. How does one penetrate that green-bladed mask, what might the nature of the river be, where does its source lie? The river flows beneath the earth, or beyond it. We have seen the whole earth from space, we have peered into space itself, we have seen the invisible with our instruments, and still the questions outweigh the answers. The meaning of life, if there is such a thing, eludes us. The

meadow contains the weight of those questions, all that man has asked and many that he has not learned how to ask, for the meadow with all its parts is merely a tiny piece of the larger whole. It is a puzzle within a puzzle, but we can only begin with the parts at hand.

A mosquito had landed unnoticed on the back of my wrist, and the sting of her beak roused me from the void. She was pumping blood from my capillaries, and her abdomen was soon swollen and red. How much does she know, I wondered, about the rivers which flow beneath her? How much can she possibly understand about liver and heart, lungs and bone marrow? If I speak the magic words to her, words like *ventricle, aorta, vena cava, hemoglobin,* will the blood taste sweeter to her? In that single moment the life of a mosquito

During the spring of our second year, we begin to sort out the roles each organism plays in the web of life. The plant-eaters were everywhere—caterpillars, katydids, grasshoppers, and crane flies, whose larvae feed among the roots of grass.

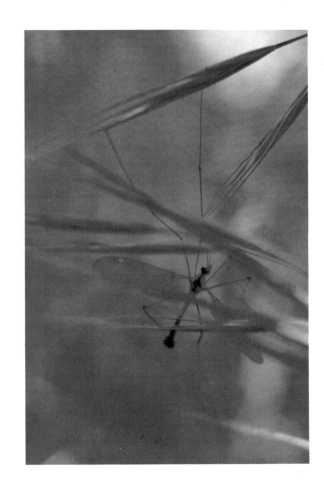

seemed infinitely simpler and easier than my own, but less exciting. I both pitied and admired her. I was torn between swatting her and letting her live, and I was suddenly terrified of my own power and my own ignorance. I, too, lay poised over a river which I could not comprehend. After all, I thought, we have at least that much in common.

I let her finish feeding. She would turn her meal into eggs and lay them in floating rafts among the sedges of the meadow bog. In due course this single drop of my blood would swim away in the reed-choked shallows, transformed into several dozen tiny wigglers. I peered through the green screen, and in the darkness of the void something moved. I thought at first it was the mosquito, but then it was gone. It had vanished, but in its place a word lingered. The word was *energy*. I scratched the welt rising on the back of my hand. It had been a fair exchange.

Our second year was splendid. We moved some distance beyond town into the country, about halfway toward our meadow. Our home was a small cabin heated by a Franklin stove. We started a large vegetable garden and, in general, lived closer to the land than we ever had before. Maggie's early years in the city, the many hours we had spent studying dried and pickled specimens in biology classes, and the long string of apartments we had once called home were only dim memories. This world was more basic, somehow more genuine, and everything seemed to function in the age-old, balanced way.

Forest-clad slopes, rolling hills speckled with oaks, and the broad meadows and pastures of the valley created an aura of immediacy, a sense of the here-and-now. We knew what phase the moon was in, when the shooting stars would bloom, when the ranchers would have to start feeding their cows hay. We were closer to the source, nearer the ancient homeland of life, planted more firmly in the bedrock systems which sustain us all—all things that share the phenomenon of being alive.

The best feature of our new residence was its proximity to our meadow and to many similar plots and slopes of grassland even closer to home. In fact, we soon found that there was much to be discovered in the fields near our cabin. Most of the year was spent nearby, for the garden called us and needed frequent attention, and the meadows called us, too, with the haunting strains of the unknown.

We were far from home, however, in the foothills of the Sierra, when we made the first new observation of our second spring. Through a mere chance of timing we witnessed the emergence of the scorpion flies.

The month was April, and the weather was warm. The grasslands were green and lush, and the blade carpets rippled in a gentle breeze. The leaf-eaters were everywhere—young katydids, caterpillars, aphids, and beetles. The fields and slopes, especially in the shade of freshly budding oaks, were teeming with crane flies—gaunt, spectral creatures with long, frail legs; they flew short distances through the grass, more like wisps of mist than parcels of living flesh.

As kids we used to call them "mosquito hawks" and fully expected them to descend upon us for a meal of blood but being interested in such things beyond the realm of myth, I soon learned from my small library of entomology books that the adult crane fly takes little, if any, food, least of all blood. Most have mouthparts suitable only for sipping water or nectar and live just a short while as mature insects.

I used to help my grandmother pull weeds in her country garden. She had a distinct distaste for grass and looked upon whole fields of it in the orchard as a toilsome invasion to be pulled or cut at any cost. My urge to help was stronger in those days than my desire to let nature take its course, and I was guilty of destroying many another animal's home. But I learned things in the process, nevertheless, among them the fact that many crane-fly species pass their larval stage among the roots of grasses. We often found them in the clumps we pulled up, and we found the pupal cases there, too, either full and ready to emerge or already empty, dry and papery. I learned from reading that most crane fly larvae are vegetarians that feed through the winter on decaying organic material. I learned, also, that expertise in such matters made little impression upon my friends, to whom the crane fly would remain, perhaps forever, a giant hawking, blood-sucking mosquito.

Maggie and I and a young friend named Jon were returning to our foothill camp from a morning walk when Jon, who has keen eyes and a quick sense for the significant, called our attention to a strange insect in the grass. I thought at first glance that it was a crane fly emerging, for it was pale and tender and had long legs. But when I looked more closely and saw that its wings, freshly expanded and still opaque, numbered four instead of two, I recognized the delicate creature as *Bittacus,* a member of a peculiar and ancient order of insects known as scorpion flies, whose ancestors have been found in fossils 200 million years old.

By a stroke of good fortune we had happened into the grasslands just when the scorpion flies were emerging, and a little searching revealed many of them hanging from grass blades by their front feet, basking in the sun and pumping fluids into their wings and legs. They were strangely pale and translucent and looked far more helpless than they would soon prove to be.

We watched them through the day. Gradually their exoskeletons hardened and their wings cleared. They began flying when we approached, and their flight pattern was scarcely less clumsy than that of the crane flies. Even so, we later discovered, *Bittacus* is a skillful hunter and, though we never observed an actual catch, by midafternoon we were finding scorpion flies with prey.

In one grove of trees we spotted a scorpion fly working over an especially large carcass. Its skin had darkened now to the rich orange characteristic of the species, and it was hanging from the grass by its front feet, much as we had observed among the new arrivals in the morning. The hind legs no longer hung limp, however, but were busily engaged in holding the prey. On close inspection we discovered that the hind feet of *Bittacus* are specially modified, like the front feet of the praying mantis, for clasping its victims.

Though the bulk of it had been consumed by the time we arrived on the scene, the insect upon which our scorpion fly was feasting was unmistakably a crane fly, and its legs dangled in disarray among those of its captor. Here, in the vise of the hunter, a tiny piece of the web of life was held. Another shadow passed in the dark river, and I remembered the mosquito and wondered if the blood I had shared with her had taken wing by now.

At least the mosquito knows something of plants—those remarkable trappers of the sun, transformers of energy, and synthesizers of all

More is held in the hunter's grip than prey alone.

When the scorpion fly first emerges, it is soft and tender. As it hangs from the grasses and its wings and exoskeleton harden in the sun, it is difficult to believe that this delicate creature is one of the fiercest predators of the grasslands.

When we found one feeding on crane flies, however, its role in the food web became apparent.

food—for as a wiggler in the meadow pond, it feeds on rotting leaves and strands of algae. Only as an adult, and only then in the case of the female, does the mosquito leave the leafy realm to feast upon blood, the sun's energy once or twice removed from the green cells where first it was held in chemical bondage.

Bittacus, on the other hand, has given up all direct use of plants to become a full-time carnivore, and though as a larva it creeps through the humus layers and grass-roots mazes of the soil, where the crane-fly larvae also reside, it consumes only flesh, devouring bits of carrion as well as small dwellers of the dark.

Through the rapidly changing months of spring, Maggie and I studied the interactions of living things in the meadows near our cabin. We became especially engrossed with that vast group of carnivores, the spiders, for they were present everywhere in large numbers and a dazzling array of forms.

There may be as many as a hundred thousand different species of spiders in the world, and grasslands are home for many of them. The number of individual spiders a single region can support is staggering. One scientist who has studied various types of grassland in England estimates that a single acre may contain two and a quarter million spiders; when one figures that on the average each one may consume at least an insect a day, and probably more, one quickly gains an appreciation of spiders as a control over the insect hordes. Imagine, then, how many insects there are in a single meadow, and how prolific they must be!

For spiders, catching insects is a chance affair, whatever the method employed, and because of this they have an amazing ability to fast for

One grassland acre may contain over two million spiders.

days, even months, and then gorge themselves when food is plentiful. However, I don't believe we ever saw a thin and starving spider in the grasslands, least of all in the spring.

We saw dozens upon dozens of spider species in our wanderings. The identification of most of them we must relegate to the refined skills of the specialists, but the broad categories we could recognize well enough.

We found stilt-legged harvestmen picking their way gingerly through the grass in the morning, at the end of their hunting rounds, or in the evening when they began again. As children we called them daddy longlegs, but then crane flies were sometimes known by that name, too. Such are the pitfalls of common names. Actually, harvestmen are not spiders, either, but represent a separate order along with spiders, scorpions, pseudoscorpions, mites, and many other groups that all belong to the large class Arachnida.

Harvestmen spin no webs. They lack silk glands entirely and must catch insects on the prowl. Female harvestmen, having never evolved silk even for the protection of their eggs, deposit their eggs carefully in soil crannies, where they soon hatch into tender white miniatures of their parents.

Usually harvestmen catch small insects, mites, millipedes, and other minutiae which they encounter during their wanderings, but some species have been observed feeding on dead animals, from worms to small vertebrates. There are even records of harvestmen feeding on bird droppings, fungi, seeds, and other bits of vegetable matter—observations which clearly place these creatures among the omnivores, where one would scarcely expect to find any relative of the spider and the scorpion. Specialists devoted to the study of harvestmen have

A harvestman, stopping to clean a leg, searches for a hiding place at dawn while, at the same time, the plodding tarantula returns to its burrow. These predators are active at night, hunting their prey "on the hoof."

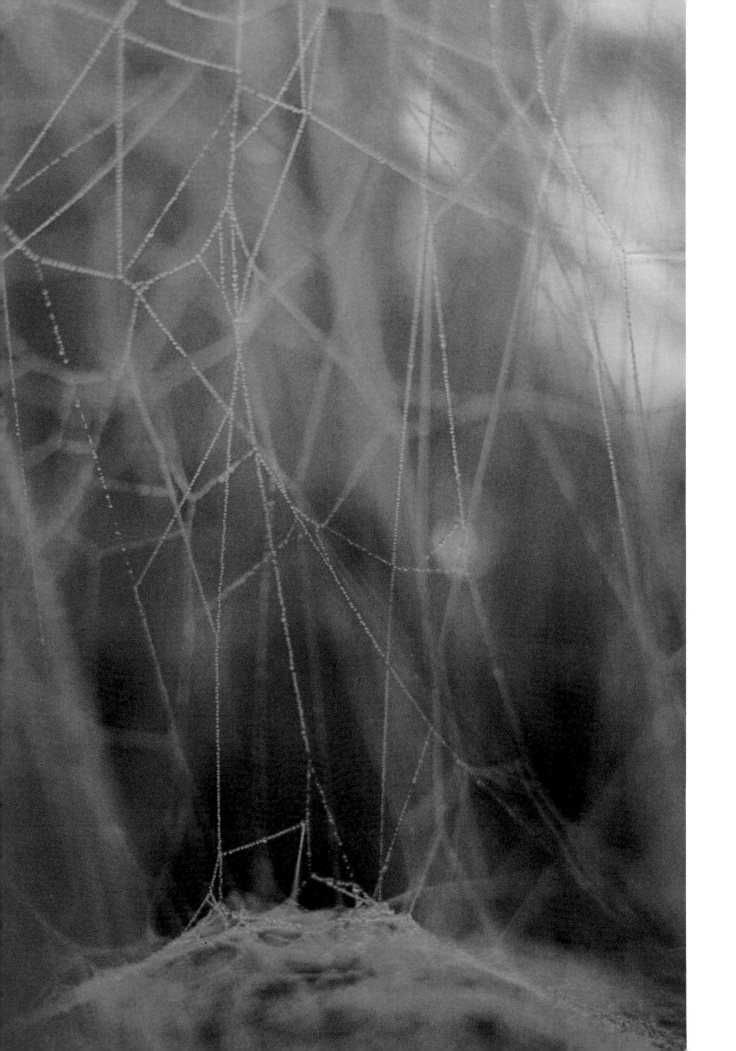

Through millions of years of evolution,
spiders have developed some of
the most spectacular architecture in the
natural world—the dome web,
the bowl-and-doily web, the snare
of the triangle spider, and the magic
creation of the orb weaver.

During our second year we followed the life
cycle of the golden garden spider. The female
wraps up her prey in a blanket of silk.
Slowly she drains the insect of its body fluids,
unmindful of small flies which come
to share her meal.

also discovered that the second pair of legs somehow aids in the recognition of food, since the eyes are poorly developed. Harvestmen lose legs readily, often as a means of escaping predators, and are perfectly at home on only four of their eight appendages. If, however, an individual loses both legs of the second pair it soon perishes.

I'm not sure whether it is stretching things to say that daddy longlegs are extremely fastidious, but during many walks through the grasslands we watched them pause in their explorations long enough to clean one or several of their threadlike legs, pulling them carefully one at a time through their mouthparts.

What is the function of this judicious preening, and what selective forces could possibly have led to its development? These and other questions loomed frequently before us, not the least perplexing of which was how the harvestman was given its name, but answers would have to wait for another time. Discoveries piled upon discoveries, and some new creature was always diverting our attention. On a late summer afternoon one such diversion was provided by a very large tarantula.

Our paths met on a grassy hillside where the oaks grew well apart from one another. Numbers of wolf spiders scurried ahead of us, startled by our crackling approach through the dry grass. Their small size must have accentuated our impressions of the tarantula when we encountered him, even though we were certainly justified; with a leg-span of three inches, the tarantula is our largest spider.

Wolf spiders and tarantulas are hunting spiders. They never evolved the ability to spin snares; instead, they have retained the ancient way of capturing their food "on the hoof." The recipient of our sudden attentions was an adult male, on the prowl either for prey or for the burrow of a female. There was a slow, relentless, almost primeval quality in his movements, as though he had been arrested in a time warp different from our own. Actually, this was not entirely a fanciful notion, for tarantulas and their kin are very little changed from ancestral forms, which were among the first animals to leave the sea and adapt to life on land.

As we watched the spider plod through the grass, leg over hairy leg, we instantly understood why tarantulas have figured prominently in science fiction and local mythology, though nearly everything that has been said about them is either false or exaggerated. While their poison is quite sufficient to subdue insects or even small frogs and lizards, no species in the United States has a bite much more efficient in its reaction on humans than the sting of our good friend the honeybee.

My favorite acquaintances in the meadows near home were the jumping spiders, if only because they evoked images from those early years when fields and "bugs" were new to me and the beady little eyes of these alert creatures seemed everywhere focused upon me. I grew to love them, as I have ever since, and when I see one skipping from blade to blade, or hiding around the back side of a stem, I can smell again those first fields and, inside among the memories, I rejoice.

Some species of American jumping spiders are brightly patterned with red, black, orange, or white, but they achieve their greatest splendor in the tropics. In Central America and Southeast Asia I have seen species elaborately marked with the most brilliant metallic and iridescent hues. One of the great tragedies of man, I remember thinking once in a steaming Malayan rain forest, is that he is blinded by his own peculiar sense of scale. I had found a gem of a jumping spider less than half an inch long, and as I studied him in my hand—and he studied me in return—I realized that if only this magnificent creature were larger he would be as familiar to us as the hummingbird, peacock, or bird of paradise.

A particularly drab individual, though, impressed me more than any of these exotic equatorial creatures, and I encountered her only a few yards from the back door of our cabin. By chance I witnessed a female jumping spider less than a quarter of an inch long tackle a crane fly six times her length and many times her weight, and subdue it completely within a few moments. Here, I thought, is a hunter of superb skill!

Like the wolf spiders, jumping spiders spin no webs for catching food; but unlike their more primitive cousins, they are specialists of the highest order. Their particular course of evolution has given rise to the finest eyesight in the class Arachnida. Four sets of eyes produce a sharp image at ten or twelve inches, and awareness of moving objects is possible at greater distances still. Jumping spiders are active by day, spying their prey at a distance then stalking slowly closer in preparation for the final leap. Unlike other spiders, their courtship is visual, involving elaborate rituals in which their patterns are displayed. This accounts for a frequency of brilliant colors unique in the spider world.

Though more secretive by nature, black widows were also frequent inhabitants of our hillside haunts. We seldom saw them—they remained hidden beneath rocks or in abandoned rodent burrows—but we soon learned to recognize their presence by their characteristic messy mazes of tough silk and, late in the summer, by numerous small males lurking expectantly in the silken tangles above retreats of the females.

It is true that black widows possess venom potent enough to make an adult quite ill, in some cases. Bites prove fatal on rare occasions, especially to small children, whose curious and probing fingers are likely to find the large females unexpectedly in a basement or beneath a rock. Yet it is equally true that the black widow is one of the few spiders in North America at all harmful to man. One of the most telling indications of how far removed our culture has become from the realities of nature lies in our blanket abhorrence of spiders.

Our predecessors on this continent, the many native peoples of America, lived in close harmony with the earth and its varied forms of life. Though their knowledge was unscientific as we would define it, they looked upon other organisms in remarkably accurate perspective. Spiders were included in many Indian legends and, by and large, were looked upon as benefactors rather than as emissaries of the devil.

The story of creation, as told by the Sia Indians, placed a spider in the position of the great wizard. This spider created two women who in turn gave rise to all living things—one to the Sia themselves and the other to all other races of men. The Pima told a different version, in which a being named Earth Doctor created the world from a small amount of dust. As he danced and sang, the world expanded beneath his feet, and he created the sky. But the earth continued to tremble and stretch until the sky did not fit properly. He then created a spider to spin together the sky and the earth so that the world would be stable.

Many tribes told tales in which Indians in danger sought help from friendly spiders, who spun ropes by which they were able to escape

American Indians looked upon spiders as benefactors.

In the fall, after months of feeding and growing,
garden spiders are ready to mate. Diminutive
males congregate, plucking special vibrations
of courtship on the web strands while
they wait for a chance to approach the female.

from their various predicaments. The Apache and Shoshoni regarded spiders as the source of their superb knowledge of weaving.

The ancient Greeks, too, had a legend about weaving. A young maiden named Arachne wove so well that she surpassed the skill of the goddess Athena. Arachne's talents at the loom so aroused the wrath of Athena that she cast the maiden in the form of a spider, thus condemning her forever to the toils of spinning. It is from this legend that the class Arachnida, to which all spiders and their relatives belong, derives its name.

In reading about spiders I found also that the native peoples of New Guinea, Papua, Queensland, the Solomons, and the Trobriand Islands have devised a wide variety of ingenious nets and lures for fishing, fashioned from natural materials and utilizing spider silk as the principal ingredient. Silk produced by spiders is the strongest known natural fiber. It has a tensile strength second only to fused quartz, and is much tougher than steel. This combined with incredible elasticity, and the quality of retaining both properties even when submerged in water, renders spider silk a potent tool in the skillful hands of Pacific peoples whose survival depends upon inventiveness with the materials immediately at hand.

The product of spiders has not been entirely overlooked in the West, either. Various efforts have been made to employ spider silk in textiles on a commercial scale, and in the United States, especially prior to World War I, the silk of several species, particularly the black widow, was used extensively as cross hairs in the optics of engineering and surveying instruments. Some of this equipment played a small role in the First World War.

A discovery made between the pages of a book can be just as arresting and potent as a discovery made from personal observation. In the spider myths and simple technologies of other peoples, I see a kind of balanced bond with the surrounding world; in the attitudes and commercial ventures of our own culture, I see a rift in the bond with nature. It is somehow fitting testimony that we have used the black widow, dreaded symbol of wilderness evils, in waging war upon our own species.

In the valley around our cabin I found a tiny spot that we had missed many times the previous year on the way to our meadow. The place was a gem, a little grassy swale tucked against the hills, surrounded by woods and bisected by an abandoned bend of the old road which had been partially reclaimed by pavement-breaking mushrooms, shoots of grass, and drifts of fallen leaves. It was an amphitheater studded with spider webs; we spent many mornings of our second year studying their varied designs.

Close to the ground we found sheet webs and funnel webs. Scattered between them was an assortment of maze webs higher in the grass and built rather on the plan (or the lack of plan, to be more accurate) of the web of the black widow, except that each had an added component of a more regulated nature. The dome-web spider suspends within its haphazard superstructure a symmetrical and finely built dome, beneath which the architect hangs upside down waiting for a meal. Another design consists of a bowl-shaped component constructed in the maze above another layer of silk, in a manner which has given its maker the appropriate common name, bowl-and-doily spider. Still another variation in this combination

Spider silk is the strongest known natural fiber.

approach to web building consists of a small orb web hung within the general tangle.

By what course of events could such a variety of web-spinning architecture have possibly evolved? They were all there in this tiny parcel of countryside—many species of spiders living side by side, yet each responding according to its kind and according to the dictate of a particular and ancient legacy of genetic messages.

Most incredible of all were the orb webs. Throughout the year we found quite a variety of them in the clearing as well as in the neighboring woods. Some were spun by tiny spiders scarcely an eighth of an inch long. The largest of all were the huge orbs of the banded and the golden garden spiders that seemed to make their appearance suddenly in the fall.

One morning I had left home early to catch a few of the slower-paced orb weavers in the final stages of their nocturnal spinning. There is hardly a more fascinating pursuit than to watch the step-by-step construction of one of these arachnid masterpieces. As I watched one banded *Argiope* methodically place her spirals, I could not restrain the thought that if every student of architecture or engineering could experience and attempt to comprehend the spinning of an orb web by a creature that lacks eyesight, intellect, creativity, and most other attributes of which we ourselves are so proud, the products of our technology might be far more functional than they are, and their creators considerably more humble.

Walking back along the old road, I caught sight of a most peculiar web, completely new to me, in the shape of a wedge. Its appearance suggested a somewhat deranged baker who had baked a single slice of pie and forgotten to make the rest. Being hungrier than curious at the moment, I snapped a couple of pictures and continued home to report my find to Maggie over breakfast.

We dug into our accumulating pile of literature on spiders and were not long in discovering that the builder of the single slice of an orb is the triangle spider, only two species of which occur in North America, one in the East and one in the West. We were happy to know its name, but the discovery did not end there. A few days later, when my slides returned from processing, I examined the two pictures of the triangle web carefully and noticed that though the twigs to which it was attached appeared in the same places in each photograph, the spider and the web itself did not. I placed one slide over the other and found that the spider and the web had indeed moved!

Further reading revealed that the triangle spider employs a remarkable method of snaring a meal. She gathers up a slack loop of the line leading to the web and holds the rest of the web taut. The instant an insect lands in the snare she releases the tension, regains it, and releases it again until her prey is hopelessly entangled. Once she has fed she must completely rebuild her web.

Our rooting around among old magazine articles also produced some notions about the evolution of spider snares. Most scientists who have studied this intriguing subject agree that silk must have developed as a modified excretory product which somehow became used as a dragline strung out behind the spider everywhere it walked. All modern spiders, whether they spin snares or not, use silk also to protect their eggs. This was perhaps the first major use to which the silk was put. The first primitive webs were probably a few strands or accumulated draglines extending to and from the burrow or hiding place, and acted by chance to

Each spider responds to an ancient legacy of genetic messages.

77

Spinning the egg cocoon is the female garden spider's final act.
Protected within several special layers of silk, the eggs
must face the rigors of winter alone. In the spring, when the
infant spiders hatch, the one-year cycle begins again.
Out of a thousand eggs, only a single pair of garden spiders
will survive to reproduce.

augment the food supply procured by direct hunting. From this crude beginning have come the maze webs, the various tangles containing more elaborate structures, and finally the miraculous orb web itself, upon which its builder is totally dependent for survival.

The most advanced orb weavers are specialists of the highest order. From over six hundred silk glands they spin five kinds of silk through numerous spinnerets and put their products to a variety of uses. Even more specialized forms have lost the ability to spin webs of their own and instead live in association with the webs of other species. The jumping spider represents the current apex of another line of development, while the triangle spider belongs to still a different group which is thought to have diverged very early from the mainstream of spider evolution. The triangle spider and its allies spin silk in a fashion quite different from other spiders and must have arrived at web spinning independently.

These pathways of spider evolution can only be surmised. They are studious guesses based upon web structure, spinning behavior, the position each species assumes in relation to its web, and the anatomy of the spiders themselves. What sequence of mutations, what incredible series of selective events actually took place to bring forth through millions upon millions of generations the spiders we know today will possibly forever remain unknown. Life moves through time like a fleeting shadow, flowing, changing, and leaving few traces of the places it has been and the forms it has taken. In this quality of life I find my deepest source of fascination. I find here also my greatest apprehension, for I must conclude that since man, too, is a living species, he is part of the same elusive shadow whose beginning lies in

Life moves through time like a fleeting shadow.

the void of the past and whose end cannot be foretold.

All the large garden spiders we saw in the fall were adult females feasting on their last insect meals before winter closed in upon them. They were storing up for the great, culminating event in their lives, the production of eggs. We rediscovered in our reading what we had observed as children, that the male *Argiope* spiders are much smaller than the females. They spin orb webs throughout their youth, but once they shed their skins for the last time and emerge as sexually mature adults they lose the ability to make snares and begin a nomadic search for the females. During our autumn walks we watched for their appearance around the large webs of their mates, and we were soon rewarded.

Golden garden spiders were abundant in the coyote brush bordering the old road. In their webs we found our first male, and before many mornings had passed, each female was attended by one or more of the slender dwarfs. In one web alone we counted eight males!

For most male spiders life during mating season is a precarious affair. The females throughout their lives have attacked for food any creature that moved within their grasp. For these web-spinners, anything that entered their snares was quickly paralyzed and drained of its vital fluids. How, then, can a male spider approach his mate without being attacked and killed before the essential act of copulation is consummated?

Spiders have no social organization, as do wolves or lions, nor do they choose lifelong mates, as do many birds. By and large, they are loners. Yet their mating problems have been solved by adopting the same basic mechanism that works in one form or another for all predatory animals—communication.

Hunting spiders have developed a variety of courtship rituals through which the male identifies himself before the female so that she will accept his advances. The tiny male garden spider, whose mate is blind, must by necessity communicate with the female in the only terms to which she can respond: the suitor must vibrate the web strands in a manner that distinguishes him from potential prey. In spite of his web plucking, he approaches the mammoth female cautiously, keeping his distance until the time is right. At best the female pays little attention to him, sometimes brushing him aside with the flick of a leg after mating. Once in a while, in her irritation at the intrusion, the female kills and consumes her consort. Biologically this is of little consequence because once mating has been accomplished the male soon dies anyhow, his lifework completed.

Life is short for the female, too. Unlike the female tarantula, which may live for twenty or thirty years, most spiders, including *Argiope*, live and die within a single year. Not long after her eggs have been fertilized, the golden garden spider begins her final act: the laying of her eggs and the spinning of a protective cocoon to house them.

The skill with which she spins the egg cocoon is every bit as remarkable as the production of the glorious orbs she spins during the rest of the year. She chooses a protected spot deep within a clump of grass, or the branches of a small shrub. Here she spends hour upon hour at her work. When she has finished, her energy is entirely spent. Shrunken and exhausted, she soon dies.

What she leaves behind is the future of her species, at least her small part of it, and she provides for it exceedingly well. The glistening brown, tear-shaped cocoon contains roughly a thousand eggs. They are protected by three layers of silk—a thin white inner lining, a thick fluffy blanket, and a tough waterproof outer covering. There are six components in all, and within this envelope the eggs will pass the frosts and storms of winter. In the early spring the eggs will hatch. After a while the young spiders chew their way to freedom. They remain together near the abandoned cocoon for a time until they are ready to disperse, settle down, and start spinning little orb webs no bigger than a quarter.

During our first year of exploring the grasslands, the garden spiders seemed to appear only in the fall, full-grown and spinning their enormous webs in the grass. Of course they were there all the time, we realized, but no dew graced their webs in the summertime to make them visible. It became equally apparent during our second year that if each female laid a thousand eggs, demise must have befallen most of them between spring and autumn.

Spiders, like all living things, are subject to many natural controls upon their numbers. Through poor placement or bad storms, some of the egg cocoons fail the test of winter. Occasionally parasitic wasps or peculiar smallheaded flies penetrate the protective cocoons and deposit their own eggs within. All through the spring and summer the growing spiders fall victim to other predators, including frogs, toads, lizards, shrews, and birds. Many are sought by hunting wasps such as mud daubers or pompilids—relatives of the tarantula hawk—searching for provisions for their larval chambers. Some are even eaten by other spiders.

The spider web, with its grand design and

Spiders, like all living things, are subject to many natural controls.

Every species faces controls upon its numbers. The dung fly, which is itself a predator limiting the populations of smaller insects, falls victim to a highly specialized parasitic fungus in one of the most intriguing relationships that has arisen in nature.

ancient history, is only a tiny portion of the living web which envelopes the earth. We can watch a spider spin its ordered creation in a single night; it took over three billion years to spin the web of life. Out of a thousand eggs only one pair of golden garden spiders survives to reproduce. Nine hundred ninety-eight spiders feed the frogs, the lizards, the birds, and the hunting wasps. This is the process of natural selection, for the web of life is still being spun.

Our second year was definitely a spider year. So many questions lured us, so many potential investigations became apparent to us that we might easily be studying spiders still, and for the rest of our days. But many more patterns were becoming visible in the fabric of the meadows; we could not ignore them. The tug and pull of inquisitiveness kept us moving. Curiosity has an elastic quality; when it takes possession, the mind stretches and weary legs are hard pressed to keep pace.

Early in the spring we made an observation that led us on another adventure into the complexities of the natural world. What we found made spider webs seem almost simple.

There is a beautiful fly, about half an inch long, that makes its appearance in the meadows early in the spring when the grass is green and the mornings are damp. The fly itself is a predator with habits similar to those of the robber fly, darting here and there after small insects. Its larvae, however, feed on dung, and for this reason the fuzzy, rust-colored adults are found most often in the vicinity of pastures where "cow pies," as we used to call them, are an abundant commodity.

As the weeks passed, we found the flies with increasing frequency firmly attached to stalks and blades, legs grasping them tightly, wings spread, yet not a bit of life remaining in them. When we examined the carcasses closely, there appeared to be a kind of mold protruding from the soft membranes between the body segments. What surprised us even more was the fact that males of the same species would occasionally swoop down upon the dead flies and attempt to mate with them!

Summer waxed and waned, and the dung flies disappeared for the year. In their place came a long string of other species, the most noticeable and annoying of which were the houseflies that collected in some quantity within our country cabin. Fall brought the first rains; a chill and dampness penetrated our living quarters, and the Franklin stove was stoked for the first time in months.

The house flies persisted. We were about to invest in some good old-fashioned flypaper when we noticed that some of our pesky visitors were dying on the walls and windows in the same fashion as the dung flies had died in the meadows. Around those that had died on the window glass, we observed white halos of powdery fungus spores, and, sure enough, we soon discovered healthy flies trying to mate with the infected carcasses. Within a week or so the fly population dwindled; the disease had taken its toll.

It was fairly obvious that we had in our midst a parasitic fungus which used flies as its host and which, when active, was exceedingly effective in population control. We wanted to know more, however, about this peculiar relationship, so we began poking around and asking questions.

A friend doing graduate work at the University of California found a paper written by a Canadian insect pathologist on the subject of a

It took over three billion years to spin the web of life.

family of fungi known as Entomophthoraceae, many members of which are parasitic on insects. We discovered, upon reading, that the genus *Entomophthora* contains some fifty species in the United States alone, which attack a wide variety of insects including numerous pests of considerable economic importance.

The actual biology of the fungus is complex, but the basic story matched our own observations. The fungus grows within the body of the host until it virtually fills the body cavities. As the insect sickens and dies, fine rhizoids from the fungus itself emerge from the ventral surface of the host and attach the carcass firmly to the base on which it rests. Shortly thereafter, spore-producing bodies emerge through the insect's exoskeleton, forcibly ejecting a cloud of spores, which settle in the characteristic halo we had seen on the cabin windows surrounding the diseased houseflies.

The fungus spreads through the population, partly by random contact of healthy insects with spores in their environment. Humidity and a certain range of temperature seem to be important in the successful germination of the fungus. This would perhaps explain that we observed the disease in the spring and fall, when cool temperatures and moisture prevail. There are records pertaining to some species of the fungus in which healthy insects have been seen feeding upon stricken ones, thus becoming infected with the fungus themselves, but I have yet to find scientific reference to the peculiar mating behavior of our flies.

Upon closer examination we could see a definite similarity between the posture female flies employ to invite copulation and the posture assumed by the dead flies. Could it be that through natural selection the fungus has developed an ability to affect its host in this bizarre way? There would probably be a selective advantage, in that the fungus would be transmitted more directly through the host population, but the question of how such behavior might have evolved staggers the imagination.

If we thought we had a specialist in our insect parasite, we were even more astonished to read about the habits of its relatives. There are five other genera in this particular fungus family, one of which also is associated with insects; members of another genus are parasitic on a group of fresh-water algae; those of still another genus parasitize the tiny gametophytes of ferns; a fourth group lives upon other fungi, and the fifth lives exclusively upon the excrement of frogs and lizards.

Consider for a moment how a fungus must locate the feces of frogs and lizards in order to survive. Apparently the spores produced by the fungus are consumed by various beetles which find and feed upon the excrement. The beetles in turn may eventually be devoured by frogs or lizards. The spores remain dormant within these animals until eliminated with the feces, at which time they germinate again as the fungus. It is a risky matter of chance, but apparently completed frequently enough to ensure survival.

I thought back to the mosquito which had taken a drop of my blood and, through the miracle of her own particular physiology, transformed that blood into a raft of eggs deposited on the surface of the meadow pond. How simple the affairs of a mosquito seemed now compared to the requirements of these specialized fungi, and yet there is room enough in the living fabric, and there has been time enough—millions upon millions of years—for such relationships to develop. I began to realize that if I were writing science fiction, my wildest inspiration could not produce as incredible a

Everything that lives becomes food for something else.

story as that which was being written right before my eyes on the cabin window.

We were much more conscious of the intricate food webs of the grasslands during our second year. Everything that lives, we realized, sooner or later becomes food for something else, but we were able to catch only tiny parts of the story.

On one hike, for example, we watched snipe flies darting after small insects. Among the other creatures we saw, a garter snake, several tree frogs, and a sparrow hawk remain in my memory. Together, they might have produced one small sequence of events in the immense web of life that prevailed in that particular meadow. The snipe fly might have captured a fungus gnat, whose larva had fed upon a mushroom, which in turn had lived upon the leaf mold beneath an oak at the edge of the meadow. The snipe fly might easily have been consumed by a tree frog, the frog by the garter snake, and the snake by the hawk.

What actually happened to those particular organisms no one will ever know, but the basic role each plays in the scheme of things we must all understand. Man has emerged as the dominant creature within that scheme. We have developed and assumed a power over other forms of life that is unprecedented in the history of the earth. Still, we are inextricably a part of the living system, and the degree to which we understand how that system functions and how our actions affect it will determine the conditions of our mutual survival.

Two great kingdoms of living things have evolved on earth—the producers and the consumers. Both need the same basic components in order to exist at all: inorganic compounds, which have their source in the earth and the atmosphere, and energy, the only available source of which is the sun. The single great difference between them is that green plants, the producers, through the unique process of photosynthesis, can trap the sun's energy and incorporate it in the production of high-energy organic compounds. These compounds produced by plants in turn become the only source of energy which animal consumers can utilize.

Animals have adopted a variety of specific approaches to their larger role as consumers. We had other glimpses of this variety during our second year. We saw grasshoppers and caterpillars, aphids and spittlebugs feasting on an assortment of green things growing about our meadows. We watched deer browsing along field margins and saw plants disappear into the burrows of gophers. We found pollywogs feeding on old leaves in the marsh pond and later watched them, as frogs, catching insects among the grass blades.

One late summer morning I was exploring the small meadow between the old road and the new one when I came upon the freshly killed carcass of a deer. It was young, a fawn only a few months old. Half a century ago it might have been the victim of a mountain lion hunting to feed its family. I remembered the eerie cries of distant coyote in the foothills of the Sierra, and a lonely feeling welled up in me. The coyote were gone. With them went the mountain lion, king of our North American predators. No longer were coyote and mountain lions controlling the number of deer, keeping the balance in the ancient way. They had been replaced by a single predatory species, man, standing behind the barrel of a gun or, in the case of this fawn, sitting behind the wheel of an automobile. The number of deer is still

Man's understanding will determine the conditions of his survival.

*How many insects does the snipe fly consume
in a single day? How many flies feed the tree frogs,
how many frogs feed the garter snake, how many
snakes feed the hawk? How many sparrow hawks
can a single meadow support during
nesting season?*

In the answers to these questions lies
one of the most basic principles of life—
the pyramid of numbers.

kept somewhat in check; I was not troubled that the fawn had been killed. I could only ask myself if man would be able to bear this new responsibility as well as the mountain lion had, and the weight of that question rests heavily on me still.

Overhead a group of vultures circled, waiting for the flesh of the fawn to soften enough for their weak beaks to penetrate. I looked closely at a wound on the hind leg and saw that several yellow jackets were cutting out pieces of meat, rising clumsily into the air with their burdens, and flying to a nest burrow in the ground a short distance away. Within a few weeks, I knew, most of the flesh would be gone from the bones.

On one of the first frosty mornings of fall I returned to the fawn. It had been reduced to a skeleton, and the whitening bones were made whiter by the frost. I lifted up the remains and found a number of carrion beetles, cold and motionless, waiting for the warmth of the sun. Some of the bones had been gnawed upon by rodents, and the little scrape marks made by their teeth were outlined in frost. I checked the yellow-jacket nest. Winter was approaching; their season, too, had drawn to a close. Some creature, possibly a fox or a raccoon, had dug out the contents and consumed the last of the workers and larvae.

The deer eats only leaves. It is an herbivore, just as an aphid is, only on a different scale. When the mountain lion brings down a sick, aged deer or a weak fawn, he plays the role of primary carnivore. When the frog eats a snipe fly, which in turn has preyed upon other insects, he acts in the role of a secondary carnivore. The raccoon which fishes for crayfish in the streams and eats berries in the fall is an omnivore; so is man. When yellow jackets kill caterpillars to feed their young, they are acting as carnivores. When I saw them removing flesh from the carcass of the fawn, however, they were filling another spot in the food web; they were the first scavengers present on the scene. They were soon followed by fly maggots, vultures, carrion beetles, and even the normally herbivorous rodents, who scraped the bones for an extra dose of minerals.

Through these basic pathways nutrients and water from the soil and carbon dioxide from the air are transformed by the plants and passed to the animals, from one organism to another. It takes a certain number of leaves to feed a caterpillar, many caterpillars to feed a frog; a garter snake eats a lot of frogs in a lifetime, and a hawk consumes many snakes. As the organic compounds travel from organism to organism, the energy from the sun travels with them. But the very process of living requires energy. A frog uses energy to croak its mating call, to leap for an insect; a mountain lion uses energy to keep warm, a lizard to crawl where the sun will warm him. At each step along the way, energy is lost. From these relationships emerges the pyramid of numbers. In the wide, green

Man has replaced the coyote and the mountain lion as principal predator of the deer. Will he be able to bear this new responsibility as well as they?

Neither man nor mountain lion
is exempt from the food-gathering
missions of the mosquito.

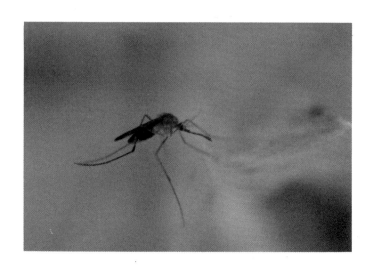

While exploring a meadow near the road,
I found the carcass of a fawn freshly killed by
an automobile. Already the first scavengers
had arrived; yellow jackets were taking small bits
of meat to a nearby nest in the ground.

On the first frosty morning of fall I returned
to the fawn. Flies, vultures, and carrion
beetles had come and gone. The final link in the
cycle of life and death would, in time, be
provided by the decomposers, that vast assemblage
of fungi and bacteria that mark the end of
the line—and the beginning.

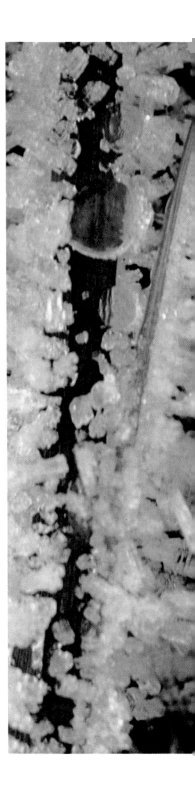

On the coldest morning of the year, we found
strange patches of super-frost scattered across
the white meadow. Frozen mouse breath!
Beneath each cluster of ice crystals was the
hidden entrance to a burrow. By the very act
of breathing, all the mice in the meadow
had revealed to us their presence.

meadow beyond our cabin, there were more insects than we could begin to imagine. Frogs were numerous, snakes few. That one large expanse, where billions of organisms lived, was the feeding territory of but a single pair of sparrow hawks.

The final link in the web of life is provided by the decomposers, that vast assemblage of fungi and bacteria that mark the end of the line, and the beginning. They inherit the last castoffs —the heaps of dead leaves that escaped the caterpillars, the deer, and the scavengers in the soil; the final remains of the fawn; the last bits of everything that was once alive. They break the last chemical bonds, use up their energy, and return to the earth all the inorganic compounds that provide the substance of living things. They are the original recyclers; without them, and without the sun, life could not exist.

Life can be boiled down to three components. One consists of the cycles themselves, the continuously repeated journeys of water, nutrients, carbon, oxygen, and nitrogen. Another is the energy from the sun, slowed in its travels through space by living things. When sunrays strike a bare rock, the energy that is

A year before, we had seen poppies only as fields of color; now we saw them as individual blossoms and as partners in age-old insect alliances.

absorbed quickly dissipates before the following night has ended. When sunlight strikes a tree, the energy might not be released until a hundred years have passed and we burn a log in our fireplace, or a hundred years beyond that when the last scrap of cellulose is consumed by bacteria. Whatever the pathways, the sun's energy is delayed in its dissipation. The greater the diversity of life on earth, the slower is this loss of energy. Whenever we pave over a meadow, drown a green valley, plant a field in a single crop, or exterminate a species, we destroy by a small amount what has taken life three billion years to build.

The third component of life cannot be seen and cannot be measured. It is the very quality of being alive. We recognize only the organic forms that contain it; as yet, the essence itself escapes our understanding. We recognize, too, that the moon, our nearest neighbor in space, does not possess this thing called life, though the sun shines upon it and the same basic elements are there. This should be reason enough to treat life gently and with caution.

Quiet mornings spent alone in the meadows are important times for me. There are things happening in the human brain that are held back, pushed down beneath the surface clutter imposed upon its circuits by day-to-day living. They are mental events that come forth only when the mind is at peace and the consciousness is free and finely tuned. In the silence of the subconscious these happenings are born; in the silence of our surroundings they find expression.

I was alone and absorbed in a casual study of an orb weaver wrapping up a fly which had just flown into her snare. I had seen it a hundred

Diversity of living things slows energy's journey.

As flowering plants evolved, a rich, concentrated
form of food made its appearance on earth.
Now wherever there are seeds, there are birds,
mammals, and insects to feed upon them. Early
in the fall, when dandelions and thistles
mature, there is a great feast in every meadow.

times before, but this time my mind was grappling with a new thought, a sort of mental game involving the web itself.

I don't pretend to be a mathematician. I am weak in my comprehension of math, even to the extent that I have trouble making the checkbook balance. Yet I found myself pondering what would happen if one started in the center of an orb web and followed first one and then another pathway along the web strands, always coming back to center again. This particular web consisted of an outer support grid within which were stretched no less than thirty-two spokes and thirty-five complete turns of the spiral. So many directions, both long and short, so many zigs and zags became apparent that my mind reeled at what seemed to be an astronomical number of possibilities.

Though I didn't pursue it very far, I could see that playing this game with the web strands would be like trying to fathom all the possible journeys that occur in the web of life itself. The hub of the orb is the source of the basic stuff of life, the first loop of the spiral is the kingdom of green plants that piece the constituents together. The rest of the web, traveled in an almost infinite number of ways, represents the routes by which the basic molecules find their way outward through the animal kingdom and back again to the source through the work of the decomposers. The deer, the vulture, the snipe fly, the yellow jacket, and all the plants that nurtured them must be there somewhere along the spokes and spiral of the orb. The frog is there, too, and the fungus whose special function has become the decomposing of the frog's excrement.

It is a puzzle without a foreseeable end, for even though the number of extant species of living things on earth is finite, time is infinite;

In the discovery
process
lies an element
of the unexpected.

the species are constantly changing, and so their ways of interacting. In the middle of the web rests the spider, or the sun, the creator who spins by strand of silk, or ray of light, each tiny part of the larger whole.

I left the meadow light-headed with the thrill of synthesis. It was not that I felt my discovery was new to man's experience; the idea was important only because it was new to me, and because it had possessed me as though, in this quiet and beautiful place, I had been put under a spell.

Several months later I read some American Indian spider legends, and I understood more clearly my experience that morning in the silence of the meadow.

The most captivating quality of the discovery process is the element of the unexpected. The suspense emanating from the unknown and the expectation of surprise are the lures that every explorer follows, no matter how long or short his journey.

Imagine how the sailors of long ago must have felt when they first caught sight of a new coast; the explorers of the West when they first saw the Great Salt Lake, the distant Sierra Nevada, and from the crest of the last range, the Pacific Ocean itself. Is the feeling any less intense for a resident of the Midwest plains on his first trip to the seashore, or for a child from the city when he first sees a mountain range, or a colt, or a cow giving milk?

When you found your first pollywog, would it have mattered whether you were exploring the shore of a duck pond in New York's Central Park or an alpine lake in the beautiful Yosemite high country? If you climbed Mount Everest tomorrow, would the thrill you experience be greater than the excitement of first reaching the top of the highest hill in the neighborhood?

On my twenty-first birthday I was exploring a nearly unknown summit in the Galapagos Islands when I discovered among the lichen-covered crags an obscure order of insects, which had never before been recorded in the archipelago. My surprise and excitement was no greater then, in that strange and exotic place, than it was several years later when Maggie and I discovered the "mystery frost" in a meadow near our cabin.

The nights had been colder than I had ever experienced in coastal California. While the temperatures were mild for winter in most of the country, in our area the time quickly became known as the Big Freeze. The frost marched right down through the salt marshes to the edge of San Francisco Bay, and the vineyards of the famous Napa Valley suffered more damage than they had in many decades. Night after night the grape growers burned smudge pots, and when they ran out of fuel, they burned rubber tires until even the birds turned black from soot. It was a troubled time, and the grasslands lay crisp and white beneath the frost.

In our meadow even the small creek had frozen. Along its banks moisture had been extruded by the cold as miniature columns of frost heave, and the grass crunched beneath our feet. What perplexed us more than anything, however, were small clusters of extra-long ice crystals scattered here and there among the grass blades. For some unknown reason, certain tiny patches of the meadow had grown a sort of super-frost; ponder as we might, we could not think of a plausible explanation.

By a mere kick of my boot, the startling answer was revealed. I had by chance laid aside the grass, exposing the entrance to a burrow. Somewhere in a cozy subterranean chamber there was a meadow mouse, whose warm, moisture-laden breath, rising during the night through the tunnel, had condensed as an extra-heavy frost on the grass at the mouth of the burrow. We had discovered, for the first time within our experience, frozen rodent breath!

The meadow was speckled with these snowy little patches. As I look back now, it seems as though every square yard contained at least one burrow entrance. But we were too cold to think. The warmth of the Franklin stove tempted us and we returned home early. Unfortunately, however, we missed a rare opportunity: counting the meadow mice in our little section of grassland.

Meadow mice, or voles, are vegetarians. Like most rodents, they form a vital link in the food web, providing an important part of the diet of hawks, owls, snakes, foxes, bobcats, and other predators. In order to survive at all, they produce large numbers of offspring, wear superb camouflage, and move cautiously and secretively. They build feeding runs beneath the grass and conceal their burrows against the sharp eyes constantly searching in their direction. Under normal circumstances, assessing the total number of field mice in an acre of grassland would be a laborious process indeed.

Yet on this particular morning, by a fluke of physics quite beyond their control, the screen of privacy normally surrounding their lives had been removed. The essential act of breathing had revealed the location of probably every mouse in the meadow. And we made no count! I can only report that there are far more voles in an acre than one would ever, at first glance, suspect. Even learning this much about the

Rodents form a vital link in the food web.

101

Like tiny ballerinas dancing, millions of thistle seeds—more than the hungry animal hordes could ever consume—are borne aloft by the gentle breezes of autumn.

food web of the grassland was something, I suppose, on a morning cold enough to freeze creeks, water pipes, and mouse breath.

Looking back on our second year, I remember events, not by the sequence of their occurrence, but by the significance of the discoveries themselves. We had spent a year becoming acquainted with seasonal change; now we were spending another year unraveling, here and there where we could, the weave of grassland life itself. What we continued to gain beyond this understanding, however, was an awareness of our own awareness.

Long before the meadow-mouse discovery, when spring was still brightly painted across every hill and valley, we went exploring among fields of wildflowers near our cabin. Buttercups were still blooming in the meadows, and California poppies were creating the first orange splashes of the season on the faces of rocky hills. It was midmorning and the day promised to be warm. The poppy flowers were just opening to greet the sun.

We roamed back and forth through the fields, attracted by the brilliant displays of color. We soon understood, however, that the colors had not evolved for our enjoyment, though enjoy them we did. They served another function entirely. The buttercups answered the sun, throwing back yellow light waves for all to see that could, and the baby blue-eyes threw back blue ones. These selective reflections bounced back from thousands of petal surfaces waving like banners in the breeze —orange from the poppies, pink from the mallows, white from the milk-maids blooming at the edge of the woods.

To the visual displays were added the sweet aroma of lupine and many more subtle scents that we were able to detect at closer range.

Finally, we crawled through the fields on all fours, peering closely and breathing deeply. Within each flower swollen anthers were laden with pollen; the colors and the odors gave their signals, and the insects came, maintaining their side of an alliance as old as the history of flowers, bees, and butterflies.

It was among the poppies that we saw something especially new. They are bold flowers. During our first year we never quite got past the blinding masses. Now we saw them as individual blossoms, as petals, sepals, stamens, and pistils, as purveyors of pollen to fertilize and ovaries to be fertilized, and as a food source for the insects which effect that vital union. We had seen small wild bees visiting poppies, but on this particular day we found, in addition, the presence of small beetles among the anthers.

They must have been there the year before, but our awareness had not yet become broad enough to take them in. I have learned since that these insects are a species in a group known as sap beetles, most members of which are found where plant juices are fermenting, especially in rotting fruit or flowing sap. Some are associated with fungi, and others feed on dried animal carcasses. That some species feed on pollen and nectar is testament to the diversity with which the process of evolution has endowed a single group of organisms.

Nearly every poppy flower contained at least one beetle, and some harbored several. What surprised us even more, though, than the presence of beetles as possible pollinating agents was our discovery in a number of the blossoms of tiny moth caterpillars, living in silk-fastened rolls of the petals and feeding exclusively on the petals themselves! Their skin was so thin and their bodies so translucent that they appeared orange from within as backlight from the sun illuminated the colorful meals they were digesting.

Many caterpillars require specialty fare: the pipe-vine swallowtail butterfly, through some peculiar means of chemical identification, seeks out the Dutchman's pipe vine upon which to lay its eggs; the monarch finds the milkweed. Their larvae will not feed on any other plants. In the golden world of the poppy there apparently lives another culinary specialist, and we had only to crawl through the fields on our hands and knees to find it.

When farewell-to-spring replaced with pink the varied colors on the hillsides and tarweed shoots began to lengthen among browning grasses, we spent time with seeds as we had the previous year, but we saw them in a different perspective. Seeds are tiny, encapsulated plants in a state of dormancy; they ensure survival when growth is impossible. Because of the nature of sexual reproduction, they also contain the promise of genetic variation—the essential accumulation of changes among which natural selection will pick and choose.

We had gained a little sense of the significance of seeds during our introduction to the grasslands. Now we were ready, in addition, to see that just as flowers and insects have evolved side by side, so seeds have developed in conjunction with their own alliances. The processes of mutation and natural selection, for example, could only have worked together to produce burrs and barbs once something was present in the environment to render them useful. Not until mammals gained some dominance in the animal kingdom was seed dispersal by hitchhiking possible.

By and large, seeds themselves are static. If a plant species is to spread and inhabit new areas, its seeds must move some distance, even if that distance is measured only in inches. For every primary agent of motion that exists on earth, there are seeds that have been endowed with an ability to move by it. There are seeds with floats for moving upon the water, with hooks and spines for traveling in the fur of mammals, with wings and parachutes for riding on the wind. Some have edible attachments to attract insects, while others, encased in tasty flesh, are consumed by birds and mammals, and excreted later along some wayward journey. There are seeds, too, with no elaborate structures at all, but neither have they been overlooked by the necessity to move; they are round and smooth so that they bounce readily and roll by the force of gravity.

I was walking to the village post office one morning, gazing blankly at the pavement and thinking about very little, when suddenly a strange texture came into focus. The roadway was covered with a nearly solid layer of seeds. Instinctively, I looked up. Perched on the telephone wires overhead were nearly a hundred very replete cedar waxwings. As I returned home, I saw my neighbor hovering forlornly over her Christmas-berry bush, whose branches were nearly bare.

About the same time, we noticed that along a stretch of the old road frequently claimed by gray foxes, fresh scats had been added among old dried ones. In place of the bones and fur of meadow mice, these droppings contained almost exclusively the seeds and skins of coffeeberries. The nearest California coffeeberry bushes were a quarter of a mile away at the upper edge of the woods. After the first rains the diet of the foxes changed again. Fruit season was nearly over, and the scats which appeared

Seeds travel with every primary agent of motion.

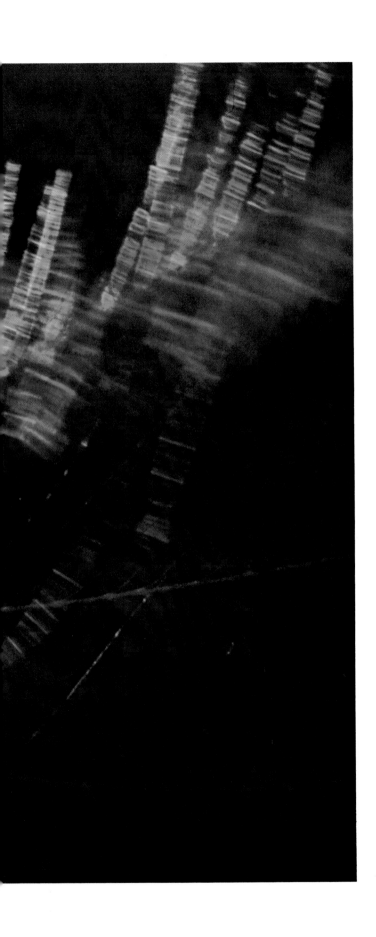

Randomly along the journey of the plumes,
as the seeds drop free, a tiny portion of
the thistle's future falls into the hands of chance.
The parachutes drift on like snowflakes, still
beautiful, but empty now of life.

upon the road contained the remains of Jerusalem crickets instead of seeds.

Fleshy coats surrounding seeds have evolved as an invitation to be eaten. The seeds pass through the digestive tract unharmed. Food stored within seeds themselves, however, is there for another reason. It is the reserve upon which the infant plant must rely between germination and the firm establishment of its roots. Seeds are the richest source of food provided by the kingdom of producers, and the consumers long ago began responding to their presence among the available supplies.

Before the rains, harvester ants were active on the dry slopes above our cabin. These small, slow-moving but hard-working insects have heavy, muscular jaws well suited to cracking seeds. Each evening just before sundown they would emerge from their nest burrows in the soil to begin their nocturnal seed-gathering safaris, and by the next morning we would find that the piles of accumulating chaff around the nest openings had grown larger.

High on one of these slopes I had discovered, earlier in the year, a few plants of a particularly beautiful, rather uncommon native species of thistle. Its silver foliage and scarlet blossom heads would make it an interesting wildflower to cultivate, I thought, and in the fall I returned to collect some seeds. As I rummaged through the dried blossom heads, I made the unhappy discovery that tiny insect larvae had been feasting on the seeds. Nearly every one had been reduced to an empty shell. I looked long and hard before I found some to take, leaving enough to ensure the survival of the population another year.

In our little swale by the old road, dandelions and other kinds of thistles had also reached the end of their season. We found middens of cracked and emptied seeds in the grass-tunnel runs of meadow mice, and small birds by the hundreds were flitting among the thistle clumps, deftly extracting and consuming the plump nutlets while discarding the plumes.

Still, the seeds forged on. Warm sun dried the ripe heads; the parachutes expanded and billowed forth. Day after day the gentle breezes of autumn were white with plumes and thistle-down, as though a soft snow were falling. Eat as they might, the consumers could not get them all. The survival of the species would not be denied.

How high might a thistle seed fly? How far might the spore of a fern or a mushroom travel upon the winds? How did insects and plants reach the Galapagos Islands across six hundred miles of open ocean, or the Hawaiian Islands across three thousand?

Many times I have watched swallows circling high up in the evening sky, so far up that they almost disappeared. I have wondered what they might be finding there. Insects, presumably, but so high? Most insects have wings, but what about spiders? Have you ever seen a spider fly?

Maggie and I did during Indian summer of our second year. They were taking off by the dozens from the fence posts around the playground of the country school in our valley. It can happen any time, but it is most frequent in the spring and fall, when tiny spiderlings hatch from their egg cocoons and the weather is calm and warm.

Most young spiders have the ability to travel through the air by a process known as ballooning. On warm days that seem to us to be nearly still, they climb to the top of some promontory and release from their spinnerets a mass of fine silken threads. When the minute updrafts rising from the sun-warmed earth tug at their lines, they release their hold and become passive aeronauts at the total mercy of the breeze. Where they will land is completely beyond their control.

Tiny insects, which are inherently weak fliers, may also become trapped in wayward updrafts and drift along as a sort of aerial plankton, riding passively on streams of air much as the floating plankton of the sea travels with the currents. In aerial trapping surveys conducted by airplane over Louisiana and other parts of the United States, insects and spiders have been netted as high as fourteen thousand feet. Inspired by the problem of insect dispersal to the Hawaiian Islands, members of the Bishop Museum staff in Honolulu have gathered in similar fashion members of sixty insect families in twelve orders at various altitudes over the Pacific Ocean. Six families of insects have been trapped five hundred to six hundred miles from the nearest land—minimum range of an air journey from the coast of Ecuador to the Galápagos Islands.

It is undoubtedly true that some organisms have reached oceanic islands by means of floating rafts of debris, and some strong fliers have arrived under their own power. But most insects, spiders, seeds, and spores have reached these outposts by passive drift upon the winds.

By the same process, insects and spiders are drawn up the slopes of mountain peaks. For the insects, such a journey means almost certain death, for they are carried far beyond their source of food into a climatic regime totally alien to them. Some of the spiders, however, have adjusted to their new circumstances. As high among the icy crags as twenty thousand feet, far above the alpine zone where the last plant life exists, there are resident species of spiders that feed upon the insects arriving with the winds from below.

As the warm days of Indian summer drew to a close, the young spiders ceased their ballooning and the last of the thistle seeds made their departure. I was walking back from an afternoon of photography along the old road when I spotted something white in the grass. I looked more closely. It was not one thing but two: a dense mass of spider silk and a thistle plume. These two structures, one from the kingdom of animals and one from the kingdom of plants, whose evolutionary paths have met under the common selective influence of the wind, these two fragile parachutes had fulfilled their functions, and their journeys had ended side by side.

Once the rains returned to California, blowing in from the Pacific in escalating waves, the thistle plume and spider balloon, and all others like them, and every fallen leaf and dried stem of grass, would be beaten to the ground, limp with moisture. There they would enter a new realm, populated with a host of organisms they had never come in contact with before. Through the work of its many inhabitants, the soil would claim them all in time.

During the heat of summer, the world of the soil sleeps. Most of the creatures who live there are so small and fragile that if you try to examine them in the palm of your hand, they dry up and die within a few moments from heat and exposure. In order to stay where it is moist, they retreat into the pores and cracks of the soil itself and wait until summer has passed.

When the first rains renew their activity, they return to the surface where they meet the organic material that rains down from above. The layer of dead plant material that covers the ground—the leaf litter beneath the spreading

In the soil death experiences its ultimate triumph.

During the warm days of Indian summer,
infant spiders climb to the tops of fence posts and
grass blades, release soft balloons of silk,
and join the wind-blown seeds in the process of
dispersal. Sooner or later, their burdens
released, the parachutes return to the earth,
silk and seed plume alike.

When the rain and dew of autumn return to the
grasslands, those thistle and spider aeronauts
that survived their journeys take hold in some
new place, a few inches or many miles
from where they began.

oaks and the mat of dead plants in the open grasslands—is the contact zone between two worlds. One is a world of light, the only place where green plants can grow; the other is a world of darkness, where scavengers and decomposers carry on their essential tasks. In this damp meeting place, this twilight graveyard, everything that was organic becomes once again inorganic. In this process, death experiences its ultimate triumph.

What are these organisms that live in darkness? Their world is as familiar to us as the feel of every footstep we take upon the ground yet as unfamiliar as the depths of the ocean. They form a community as varied and complex as the one Maggie and I had spent so long exploring above ground; the community of the soil has the same categories of consumers, the same basic pyramid of numbers, only with different characters playing the roles. After the first rains of fall, we spent days pawing through leaf litter and digging up clumps of grassland soil, in order to acquaint ourselves with some of the members of this soil community.

A few of the creatures of the soil are familiar to everyone: earthworms, the great plowmen; pill bugs and sow bugs, readily found in every garden; ground beetles one may see under rocks; centipedes and millipedes; ants of many kinds; and some termites, which, like those we had seen rising into the air for their mating flight, live more in the soil than they do in decaying wood.

These are familiar to us only because of their size. The rest of the creatures that live beneath our feet are extremely tiny. They include a host of arthropods less known to most of us than the naked-nosed wombat or the pig-footed bandicoot of far-off Australia—delicate, blind insects in the order Diplura that are so ancient

they never developed wings at all; pale white proturans less than a millimeter in length; bristletails; and varieties of tiny spiders unlike anything we had seen spinning a web in the open fields.

The two most common groups found in the soil, however, are mites and springtails. In fact, they are so numerous everywhere, from polar regions to the equator, that they are beyond doubt the most numerous legged land animals on earth. Mites are small, often microscopic, eight-legged arthropods related to spiders. Every gardener has at least encountered some of the plant-loving species of mites ravaging his crops, and much larger cousins, the ticks, should be familiar to anyone who has walked with a dog through the countryside. But have you ever seen a springtail?

Scrape through any pile of rotting leaves, dig in any plot of soil, turn over a board lying on the ground, look in a compost pile or a stack of decomposing manure, peer under a mushroom, and you will find springtails, usually in great abundance and often in a variety of sizes, shapes, and colors.

The order Collembola must contain some of the most interesting insects in the world. They are minute, totally wingless, either drab or brightly colored; and their habits, what little is known of them, are fascinating. Some have a special structure folded beneath their bodies with which they can catapult themselves considerable distances. For this ability the group has received its name. Though there are many other scavengers in the soil, springtails must account for most of the initial breakdown of organic material, and they in turn, being at the bottom of the pyramid of numbers, provide food for many of the predators which prowl through the dark passageways of their world.

Over a billion arthropods live beneath a single grassland acre.

Every organism is only the substance of life—
nutrients, carbon, oxygen, nitrogen, and water—
held in a momentary form. Sooner or later,
everything that was once alive returns to the
source for reuse. During our second year
of exploring the grasslands, we watched the rise
and fall of Wyethia. By midsummer, the spring-
perfect leaves had already been ravaged
by insects.

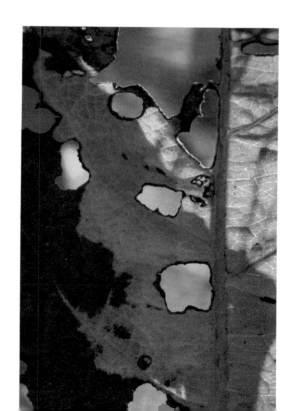

In the early fall, like a rapidly fading sunset,
the leaves briefly blushed scarlet and died.
During the winter, scavengers and decomposers
of the soil reduced them to skeletons.
On frosty mornings they seemed more like fine
pieces of lace than products of the plant kingdom.

114

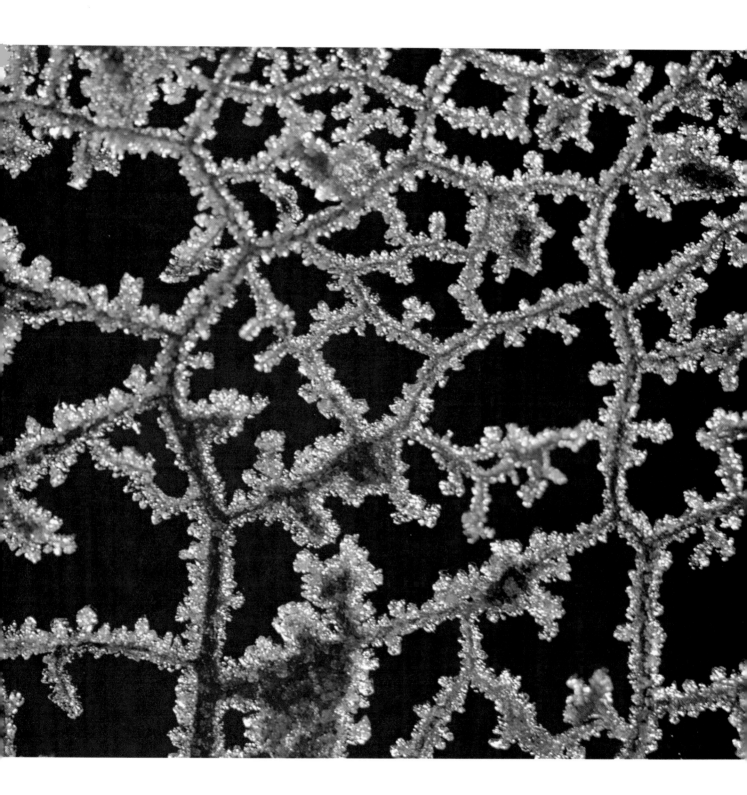

By December the new grass was up several inches.
Some components of Wyethia were no doubt
already circulating, flowing with the protoplasm,
once again a part of that intangible thing
called life.

Because of their small size, the number of soil animals that can live in a piece of earth is staggering. Scientists at Cambridge University made a comprehensive survey of soil organisms living in an ordinary English pasture. They gave the following estimates for a single acre of grassland: mites, 666,300,000; springtails, 248,375,000; root aphids and other sucking bugs, 71,850,000; bristletails, 26,775,000; centipedes and millipedes, 22,475,000; beetles, 17,825,000; miscellaneous other arthropods, 15,200,000. This amounts to well over a billion arthropods living beneath a single acre of pastureland.

To this must be added all the micro-organisms such as nematodes (round worms), protozoans, and bacteria—that vast assemblage of essential decomposers which must exist by the millions in every cubic inch of surface soil within that acre. Imagine, too, all the fungi which must send their threadlike mycelia through every stem and leaf searching for the food they are not able to manufacture themselves for lack of chlorophyll.

As we sifted through the soil, a shovelful at a time, we thought about the water we use, the food we eat, the energy and natural resources we consume, and the land taken to produce them. We wondered about the space occupied by our dwelling and the roads upon which we travel, the factories that manufactured our vehicle and refined its fuel, and the oil fields where that fuel came from. What a contrast in needs! How can anyone, at least in the biological sense, actually own an acre of land? Think of those billions of organisms whose destiny we control. How many creatures of the soil live beneath a factory? How much does man as a species, depend upon 248,375,000 springtails?

We must ask how many acres each man requires to fulfill his needs, and we must ask ourselves also how many of those needs are essential and how many superfluous. When we have settled these questions, we will know how many human beings—how many individuals of a single species—this finite globe can support and still leave room for all the other varieties of life whose healthy existence directly or indirectly sustains our own.

Soil cannot be taken lightly. Thousands of years and the work of billions upon billions of generations of living things are required for its formation. We are dependent upon it and upon the system that creates and sustains it. When it is gone from a piece of land, it is gone for a very long time, and no man can manufacture it anew. Turn over a spadeful of soil and sift carefully through it. Look closely at a handful of humus. Then look just as closely at a handful of sand and decide which bears the stamp of life. A blind man can smell the difference—we all can—and yet we continue to abuse the soil as we have throughout history.

Everyone should see a springtail.

We chose a plant at the beginning of our second year, not a single individual but one species among the many wildflowers of our local grasslands. The plant was a broad-leafed species of *Wyethia*, or mule-ears, a robust, handsome relative of the sunflower. We checked several of them off and on through the year and kept track of their changing condition.

In the early spring, new shoots emerged from the tough, perennial rootstalks. The leaves unfurled and faced the sun. By May blossom stalks appeared, and the large buds opened into sunbursts of gold. As the sun moved farther

Thousands of years are required to make soil.

Beneath the oak trees the same cycle prevails;
the substance of everything that dies sooner
or later is returned to the living. Out of the leaf
mold we found many shade-loving plants
growing and acorns sprouting. We marveled,
as everyone must sometime in his life,
that the enormous oak has such a small and
tentative beginning.

north on the horizon, the *Wyethias* set their seeds, and their leaves, tattered now with holes chewed by animals we never saw, began to yellow beneath the blistering heat and drying winds. In the early fall, like a rapidly fading sunset, they briefly blushed scarlet and died, leaving behind in their roots the extra food they had worked all season to produce.

When we next checked the *Wyethias*, the month was October. The pounding of the rain, the bacteria working in their tissues, and the tiny creatures of the soil had reduced them to skeletons. It was a cold morning late in the month, and each leaf had been carefully embroidered with frost during the night. Like the fawn, only the bones remained, and they would not linger for long.

By December the new grass was up several inches. Some components of *Wyethia* were no doubt circulating in it, flowing with the protoplasm, once again a part of that intangible thing called life.

Beneath the oak trees the same cycle prevails. Out of the leaf mold we found milk-maids growing, baby blue-eyes, and other shade-loving plants. During the winter and early spring, we found acorns sprouting also, and we marveled, as everyone must sometime in his life, that the enormous oak has such a small and tentative beginning.

What about all the acorns that ripen in the fall? Some years the oaks produce a huge crop, and yet how often do you see young trees? Do the squirrels and birds eat most of them? The Indians of California ate their share in times past. Each autumn Maggie and I go acorn gathering, usually beneath the spreading valley oaks which produce large and tasty nuts; we have developed an annual craving for acorn bread and eagerly await the fall harvest.

We noticed something about tree seeds during our wanderings. Douglas-fir and coast redwood trees produce very small seeds. They grow in damp, forested areas. Willows, cottonwoods, and alders grow in even more moist situations, adjacent to streams and ponds, and their seeds are quite minute. On the other hand, oaks and California bay trees, which prefer drier slopes, produce large, meaty seeds, and the California buckeye grows the largest seeds of any of our trees. What could be the reason for these tendencies?

A close look at a sprouting acorn reveals a possible explanation. Long before the first green shoot appears, a single very persistent taproot penetrates the ground. By the time summer arrives the seedling has a deeply established root system. That first summer is the test. Only those seedlings which have taken hold in a protected place and have their roots deep enough to find a little moisture during the long dry season will survive. The vast majority fail even then.

An oak seed needs enough stored food to

Oak trees often begin from acorns buried by rodents and birds. Even the acorn woodpeckers occasionally overlook a seed in their winter storage trees.

The first summer is an oak's test of survival.

The old, gnarled oaks are like islands
in a vast sea of grass. Lichens hang from their
limbs, caterpillars chew their leaves, gall
insects infest their twigs, birds nest among
their branches—all as dependent upon the oaks
as ocean birds are upon islands in the sea.

produce and nourish a substantial taproot, and yet this same stored food renders acorns especially attractive to animals. This seemed to us to be self-defeating until we watched the scrub jays and squirrels at work gathering bay nuts and acorns during the fall. What they didn't eat they buried in the ground, in cracks between rocks, or beneath logs. Many of these caches were forgotten; we saw them sprout later in the winter, and we compared their locations with those of young trees which had survived their first tests of summer. It became clear that seed-eating animals were allies of the oaks and bays rather than enemies, for those seeds that they had carefully "planted" had the best chance to survive.

In moister places the need to establish a deep root system is either less urgent or absent altogether. Perhaps for this reason the seeds of redwood, Douglas-fir, and willow are small. And yet there are oaks in the eastern and southern United States where summer rains tend to reduce this apparent necessity. Perhaps the animals themselves were the selective force in the evolution of large edible seeds, and these seeds in turn made it possible for the oaks to grow in California and even at the edges of deserts in the Southwest and Mexico. Even the seeds of the California buckeye, which are stringent in flavor and too large for a bird to handle, are often dragged some distance and planted by rodents. However, their enormous taproot is so vigorous that many of the seeds easily become established on their own.

We journeyed to our original meadow for a morning walk, because we had not been there for a long time and because we remembered that at the upper edge, where grassland merged with forest, there were some enormous Douglas-firs used as storage trees by colonies of acorn woodpeckers. We wanted to watch them carry out their annual harvest.

We found them noisily at work, boring fresh holes, cleaning out old ones, flying back and forth between storehouse and source, stuffing the holes with new acorns, all the while sustaining a raucous conversation that drifted across an otherwise quiet meadow. During the winter they would feast on their stores and on the numerous acorn weevil grubs that habitually infest the seeds, but when we found a few oak seedlings growing out of the holes, some high up on the trunks of the storage trees, we knew that even the efficient acorn woodpecker overlooks an occasional morsel.

We left the meadow with the question of acorn evolution burning in our minds, but it wasn't enough to divert us from another inspiration: it was time for our own annual harvest. The sweet, nutty aroma of acorn bread baking on a cold autumn day welled up in our memory, and we headed inland toward the valley oaks to do our share of nut gathering.

We had paid little attention to the oaks the first year, though we were among them almost constantly on our many trips back and forth between the coast and the foothills of the Sierra. But a year in the grassland had taught us that investigation has no boundaries, nor has a community of living things. Interactions do not end at the edge of a meadow or the base of an oak, and we set out to explore these magnificent trees more intimately.

The oaks, individual trees or groves of them, are like islands in a sea of grass, not entirely separate from their surroundings and yet unique in terms of the organisms they harbor. Their presence in the grassland greatly expands the possible living places that exist there. We discovered that among their leaves, upon their branches, in the shade beneath them, in the fallen leaves and dead limbs they have discarded, an unbelievable variety of organisms exist which could not live in the grassland otherwise. We had discovered a world within a world, and another direction for our investigations that was just as boundless as all the others we had stumbled upon.

Among the primary consumers, the most obvious were the caterpillars. The live oaks, in particular, were infested with tent caterpillars and the larvae of the California oak moth. Tent caterpillars hatch from a frothy mass of eggs deposited by the parent moth around a twig. Unlike most caterpillars, they stay together until they are full-grown and ready to spin their cocoons. Throughout this time they live in a silken tent which grows larger as they themselves grow. They leave it to forage among the leaves, and when they have stripped one branch they move to another and construct a new shelter.

Some years they appear in enormous numbers, but the increase in population never lasts long. Birds consume some, and as we soon discovered, they have their parasites as well. Many times we observed oval, white eggs on their skin, usually placed behind the head, and when we tried to raise these host caterpillars to adulthood we found spiny little flies in our rearing cages instead of moths.

The oak-moth larvae also increase to epidemic levels occasionally, and when they do, vast areas of live oaks are stripped of their leaves. When this happens near town, and the constant crunching sounds and the rain of droppings occur annoyingly close to sundecks and barbecues, the tree services with their fleets of spray trucks do a considerable business. "Save the trees" is the battle cry across the land, but few people seem to ask themselves why, if caterpillars kill trees, oaks have reached the venerable age of two centuries when man has been spraying them for less than a decade.

The oak-moth caterpillars also have their natural enemies—viral diseases, lacewing larvae, birds, or a severe winter. Hopefully, it will soon be common knowledge to everyone that the indiscriminate use of insecticides also eliminates some of the predators and parasites, and the target populations only seem to increase the more one sprays. It seems especially foolish to upset the balance of nature merely to quiet the munching hordes or prevent a caterpillar dropping or two from landing in an evening cocktail.

We found a number of other larval species working among the oak leaves, too, some quite minute, feeding only on the leaf surfaces and exposing their inner skeletons. Many of the smaller moth larvae remained hidden within tightly rolled leaves, spun together with their own silk.

Among the most interesting of the primary consumers, however, were the gall wasps. Oak galls of various kinds are familiar objects from coast to coast wherever oaks occur, and the huge "oak apples" of California are especially well known because of their size. Less well known, we have found, is the remarkable story of their formation. We had learned from reading that oak galls were formed by insects. In the fall we collected the huge "apples" and many varieties of smaller size, placed them in jars, and waited. Sooner or later all of the galls produced insects, and as we studied these tiny creatures and continued our reading, the mystery unfolded.

The gall makers, in this case, are wasps less

The use of insecticides is a dangerous game to play.

Ferns, mosses, and lichens that grow as epiphytes
on trees require a substantial foothold. The
rough-barked oaks are often densely covered with
growth whereas the madrones, which shed their
outer bark each fall, are bare.

than an eighth of an inch long, of the family Cynipidae. In the spring, the female wasp finds the particular oak species of her preference, selects a particular part of the tree, such as leaf, stem, bud, or root, and inserts within the plant tissues one or more eggs. Each time she repeats this process she injects an infinitesimal amount of a chemical substance unique to her species. This chemical induces the tree to produce about her eggs a tumorous growth, the gall itself, within which the larvae will feed and mature. What is especially remarkable is that the gall—and there are many startlingly beautiful forms —is specific in its structure to that species of wasp. To further complicate matters, some gall wasps during another time of the year will induce a different kind of gall entirely on a different part of the tree's anatomy.

The story does not end there. We raised from our collection of galls numerous other wasps which are either parasites of the gall wasps or parasites of the parasites! Some of these are so tiny that they can be seen in detail only with the aid of a microscope, and they are so splendidly marked with metallic hues that they rival the jumping spiders of equatorial Asia or the most stunning of tropical butterflies.

Once the galls fall to the ground and the occupants emerge, other organisms inhabit them. The giant oak apples, especially, become virtual apartment houses for snake-fly larvae, spiders, and a host of other creatures.

Wherever there are insects, there are birds. Because oak trees are alive with small forms of life, they attract also an assortment of birds, especially perching birds, which are associated with trees and would probably not venture far into the grasslands were the oaks not there. We identified many species, some hunting caterpillars among the outer twigs, some searching along the bark of trunk and limb where insects and spiders might be hiding in the cracks, some even pecking into the galls after their particular contents.

The oaks not only harbor a rich fauna, we discovered; they also provide living places for an array of other plants. Two such dwellers among the limbs were lichens and mistletoe.

Mistletoe prefers deciduous oaks, and in the winter great green clumps of it can be seen hanging from the bare branches, often in considerable numbers. Mistletoe is a parasite. The white, waxy berries, familiar to everyone who has hung a sprig over the door at Christmastime, are a favorite food for many birds. As they perch among the limbs after a meal, their seed-bearing droppings may land on a branch, and if a seed sticks in a favorable niche it may germinate there, sending rootlike structures directly into the tissues of the host tree. The mistletoe on oaks is green and thus manufactures its own food by photosynthesis, but the water and nutrients it uses, it must take from the tree itself. Though few trees are killed by their parasitic associates, some of the oaks we saw which were especially burdened with these plants were obviously not in good health.

Lichens, however, are not at all parasitic upon the trees. They are epiphytes; they simply use a tree as they would a rock, as a solid base to which they are attached and as a nonliving source for nutrients. Oaks growing near the coast, within range of cool, moist summer fog, were especially covered with these plants, and we often found surprising numbers of species growing upon a single tree. Most beautiful were the long pendulous strands of gray-green fishnet lichen and old-man's-beard, which

*Lichens
are two plants
living in close
association.*

128

sometimes densely draped the trees. Locally both are called Spanish moss, a name also applied to the drapery common upon trees in the South. In neither case is the plant a moss; in fact, southern Spanish moss is a flowering plant related to the pineapple!

Lichens are remarkable plants. Each species of lichen is in actuality two plants living in close association. The structures we see as the "lichen plants" are various species of highly modified cup fungi, which cannot live independent of the microscopic single-celled algae they always contain within their tissues. Many questions remain to be answered about the biology of this peculiar relationship and how it evolved, but it is thought that the fungal component of lichens is partially parasitic upon the algae, deriving food material from them while providing them with a place to live that is more moist and protected than one they might find living free in the environment. Together, they derive their nutrients from particles of organic material washed down by the rain, and by slowly breaking down dead outer bark without affecting the living tree.

We realized, too, as we studied other kinds of trees surrounding our local meadows, that the oaks were more prone than many to the growth of epiphytes upon their limbs. Besides lichens, the oaks near our cabin were often densely clothed in a thick layer of green mosses, and upon some of the older branches grew polypodium ferns. Only the Douglas-firs and broadleaf maples, of all our woodland and forest trees, seemed to compare with the oaks in this respect, and we were not long in tracing the difference to the nature of the bark itself.

Epiphytes need a substantial foothold, lichens less than mosses or ferns. The rough, corky bark of oaks, firs, and maples provides that foothold, retains moisture, collects organic material, and provides some nutrients in addition through its own decomposition. In sharp contrast is the madrone whose smooth, orange bark is always entirely free of epiphytic growth. No wonder! Every fall the madrones shed their outer bark, much like eastern birches; a thin layer curls and peels from every branch and twig, revealing a new green bark which soon acquires its magnificent color.

Winter and early spring are the best times to study epiphytes in our area. The plants are moist and active then. The polypody ferns are green with fronds; after a rain the mosses form a soft, damp carpet on the limbs of the oaks; and the lichens are pliable and elastic. Besides, there is hardly a better time of the year to be outdoors than after a heavy rain. The air is cool and clean, and the fragrance of damp earth and decaying leaves permeates the soul. We spent a lot of time in February and March with these little plants, occasionally taking some home for a closer look through the microscope.

We had been out all day on one of these moss-gathering adventures and decided to stay until sundown. We were well rewarded for our decision. As the tree-studded horizon swung toward the sun, we heard the familiar hoot of a great horned owl. We stopped and listened for the direction from which the call came. The next round placed the source in a big old live oak only a short distance away. Silently and stealthily we crept closer until we could see the huge bird perched on a limb, catching the last rays of the sun. We sank slowly into the grass and waited.

The horizon finally engulfed the scarlet sun; the owl hooted a few more times and then

An owl has been mice, seeds, earth, and clouds.

On one of our evening walks we heard
the familiar hoot of a great horned owl. Silently
and stealthily we crept closer until we could
see the huge bird perched on a limb, catching the
last rays of the sun. We sank slowly into
the grass and waited.

At sunset, the owl silently glided from its
roost. It was hunting time. We remembered the
meadow-mouse burrows hidden in the grass, and
we realized that, for the owl at least, the oak
is an integral part of the grassland.

glided silently from its roost. We watched as it swept low over the meadow and disappeared into the half-light of dusk. It was hunting time. We remembered the meadow-mouse runs hidden in the grass, and we knew instantly that, for the owl at least, the oak is an integral part of the grasslands. Just as sea birds need islands or rocky shores on which to roost and nest, so the great horned owl needs trees.

Behind the silent shadow of the owl our second year of exploration closed. A single night has no beginning and no end, exactly, because it comes gradually from dusk and, through dawn, becomes day. Neither has an owl, for its very substance has been mice and seeds, earth and rock, rivers and clouds, maybe even other planets and other stars. Our second year slipped away as silently as the owl left its tree, for the process of learning also knows no bounds, at least none that we can identify.

We don't know when learning began on earth, or with what animal, nor do we know whether, on other planets in the universe where life might exist, this strange by-product of life has ever arisen before. Neither can we be sure when learning begins for each individual human being. Is it in the womb when we first hear our mother's heartbeat? Was it many thousands of years ago when fire was first tamed, or was it when our father showed us how to strike a match? Was it when the first stone tools were fashioned, or might it have been long before that when man was still a creature of the distant future? And it is just as certain that we don't know when learning will end, for the fright and the thrill of learning is that we have no way of knowing where it will ultimately lead us, as individuals or as a species.

The skills of observation and questioning are cornerstones of learning.

When did our search in the grasslands begin? Was it that first spring when we discovered a beautiful meadow? Was it when I was five or six years old and found my first snail in a postage-stamp San Francisco backyard, or when Maggie saw her first bog sundew on a college field trip?

We could have started anywhere. Our meadow contained a stream and a marsh, and was surrounded with forest. What about these places? What about a vacant lot in the city or the duck ponds in the park? We didn't need a car to drive someplace or a camera to record what we saw. We could have been anywhere, with no further equipment than our minds and the senses that feed them.

No matter where we might have begun, we eventually would have come down to the same basic principles, the same physical and biological laws that seem to guide life and the universe within which life has evolved. If we had never seen a tree or a bird or an insect, and had only ourselves to study—the way we are built, how we evolved, what it takes to keep us alive, and what happens to us when we die—we would inevitably arrive at the same truths, as long as we were not blinded by preconceptions along the way.

The challenge is not in finding the best meadow or the wildest piece of country. It lies instead in finding within oneself the skills of observation and questioning, of using our minds again as we did when we were children, and nothing in our new, strange world could be taken for granted. Rediscovering and refining these skills is the first step toward understanding, for they are the cornerstone of learning.

Many books have been written about learning and the creative process. Man is fascinated with creativity; he has every reason to be, for though man is not the only animal capable of learning, he is unique in his ability to think beyond the realm of immediacy into regions of the abstract. Some other primates are able to solve simple problems, but as far as we can determine, we alone among all the creatures that have evolved on earth are able to see in a spider, as some American Indians did, the origin of the world; to see in war and human suffering an image like Picasso's "Guernica"; or to see in a heap of fossils and a few drab finches, as Darwin somehow managed to see, an explanation for the origin of species.

From this singular human ability have emerged the three paramount avenues of creative expression—religion, art, and science. None is quite complete without the others. Religion and art emanate from the subjective nature of man, science from the objective. They are the only ways man has of defining his place in the universe. Taken together, they are the culmination of human experience.

Whichever avenue one pursues, the learning process defines the nature of the exploration; it is the prime means of gaining experience. Only from experience does the creative moment, that instant of synthesis, emerge. Without the voyage of the *Beagle* and years of study, Darwin's profound conclusions would not have been drawn. It was from living itself—long, intense years of it—that Picasso extracted the incredible imagery of his paintings. And from generation upon generation spent close to the land, the Indians created their religions.

Only the occasional inspiration or discovery that is new to the human experience and happens to be added to its permanent records is remembered. But the same great moments happen to us all, moments when many small pieces of the puzzle slide suddenly together and form a pattern we have never seen before.

We know a superb teacher of young children who told us of one such situation she observed among her students. The class had been introduced to simple equations in math; not long after, in science, they were experimenting with simple homemade balances. One boy was eagerly working his balance when suddenly, entirely on his own and for the first time in his life, the profound association was made. "Wow," he exclaimed with wonder in his voice, "I'm working equations!"

This is the learning instant. The knowledge itself is an empty thing. Had he been told in advance that he would be working equations on his balance, the wholeness of the process would have been lost. It is one's total experience, culminating now and then in these breathtaking instants of understanding, that constitutes the full and highly rewarding process of learning.

During our exploration of the grasslands, we exposed ourselves to many lives, many interactions, many processes new to us, and we added them to what we already knew. Now and then an important connection was made. No one could make those connections for us; we couldn't even make them happen ourselves, for learning does not occur by mandate. Those splendid moments occur spontaneously.

Education is more than simply learning a body of knowledge. All the facts and information we absorb throughout life are like idle ingredients on a kitchen shelf. It is not until these ingredients are blended together that creative results and real understanding emerge. Allowing these creative moments to occur is what education should prepare us for. In this sense, our two years in the grasslands had been just a small step in a lifelong education.

From experience emerges the creative moment.

When Maggie and I began our exploration of
the grasslands, we looked upon dew as simply dew,
a magic product of the night. By the third year,
we began to see in these liquid spheres the vast realm
of physical law that underlies the natural world.

UNDERSTANDING

We learned in pursuing the pathways of knowledge
that each insect or flower or drop of water in a
meadow mirrors the universe

Each insect and flower, each blade of grass and drop of water in a meadow mirrors the universe. If we could understand everything that is manifested in nature's smaller units, if we could know every principle that shapes their form and behavior, and all the natural laws responsible for their very existence, our vision would be extended farther beyond the meadow than the reach of the most powerful telescopes yet invented.

Since the beginning of man's study of science, when we first began to shed the blindfolds of myth and dogma, we have explored new landscapes with the critical eye of inquiry. We began with the familiar, and as we developed greater understanding and more elaborate tools, we ventured farther and farther into the realm of the unfamiliar.

Mathematicians began with numbers, physicists with concepts of time, space, mass, and energy, biologists with the living things at hand. Slowly, century by century, the sciences have grown. We have journeyed, in one way or another, from the most minute subatomic particles, which lie at the threshold between matter and energy, to galaxies that may be a hundred thousand light years across. The landscapes are varied almost beyond comprehension: the interior of a cell, a cloud, the earth, the ear of a whale, the eye of a mosquito; the surface of a pollen grain, a leaf, a virus, the moon. We have projected our minds backward and forward in time, and we have postulated questions to which we almost dare not find the answers. And yet the unknown has grown larger, not smaller, because our vision is wider than it was as recently as yesterday.

The structure and behavior of the universe, however, is gradually becoming known to us. This knowledge is the product of centuries of investigation by countless individuals, and yet all the principles they have revealed may well be at work in a single meadow, so connected are all the pieces of the universe by the laws that guide the whole.

When the mosquito bit my hand and drew the blood she was to use to form her eggs, I had a glimpse beyond the biological landscape I had grown accustomed to seeing. In that instant I was aware of energy that had been generated in the distant sun, the same kind of energy as is produced by countless other stars throughout the universe. At the same time I was looking both beyond the meadow and within it, for, not only the stars but every living thing contains that energy.

Could it be true, we found ourselves asking over and over again, that the closer one looks

In the meadow
is the same
energy
as in the stars.

135

the farther one sees? How many directions can one travel from a square foot of grassland? For the third year we stepped out into our meadows and began exploring.

The closer one looks, the farther one sees.

What about a drop of water? We saw many drops as dew and rain, on spider webs and in the clouds. In one form or another, water was always with us in the meadows, even during the hottest days of summer, for water and life are intrinsically woven together. Even the driest stem of grass contained it, for water is locked within the molecular structure of the cellulose itself, left behind in the dried grass-bones of the fields. Hold a cold plate over a few burning stems of grass and you will know beyond doubt that water was in them.

Water has a history. It is the history of the universe, for as water itself, or as its components, hydrogen and oxygen, or as the particles of matter which comprise these atoms, the substance of water has traveled far and wide through space. It has been born with galaxies, heated in millions of solar furnaces, transformed into energy and reconstituted as matter. It may even have coursed in the lifestreams of other peopled worlds before it finally made its way, as a drop of rain, into the fabric of our meadow.

For all its incredible work among the living, water is not alive. It is an inorganic chemical compound, and it answers only to physical laws. Because it is essential to life, the many peculiar properties of water have shaped living things in numerous ways.

Like all liquids, water has as one of its physical properties a surface tension. We know this exists when we see a leaf floating on a pond, or a water strider skating about on its surface. This tension—or "film," as it sometimes appears to be—comes about because the molecules that comprise the liquid have a certain affinity for one another. Within a body of water this molecular force is equalized, but at the surface, where water meets air, the forces are unequal. Each surface molecule, therefore, tends to be pulled inward; the liquid strives to occupy the smallest space allowed by its own molecular structure and cohesive potential, and a tension is created because water molecules cannot break away from the surface easily. A water strider that has evolved for the situation can walk on the water without breaking the cohesion that holds the surface molecules together.

A mosquito wiggler, on the other hand, would die if it could not disrupt the surface tension. Though it lives in the water, it must breathe air directly, and its siphon is so designed that it can be thrust between the surface molecules and opened to the air itself.

Though many insects and other small organisms have developed ways of coping with surface tension, either from above or below, the surface of water remains a dangerous place for most small creatures. In addition to their attraction to each other, water molecules are also attracted to most solids. When something becomes wet, this property of adhesion is taking place.

While larger animals can easily break the surface tension, small insects may become trapped by it. When a man emerges wet from a swim, he may weigh just barely 1 percent more than he did when he was dry; when a fly breaks through the surface and becomes wet, it weighs twice as much as it did before and may not be able to extricate itself from this extra burden. Surface-feeding fish and the fly-tying and tackle industry have both evolved as a result of this common eventuality.

These same cohesive and adhesive properties of water are responsible for capillarity—the ability of water to rise within tubular structures; the narrower the tube the higher the rise. By capillary action water can rise through narrow pores and passageways of the soil, providing a moist environment for organisms that live there, even though moisture continually evaporates from the surface. Capillarity has also helped to shape the evolution of conducting vessels in plants and animals. The pressure of water passing through the membranes of root hairs by osmosis, its capillary movement through the vessels, and the "pull" of evaporation from the leaves work together to transport water from the deepest roots to the highest twigs of a tree; the physical capabilities of this system effectively limit the maximum height to which trees can grow.

One morning during our third year, Maggie and I were walking very near our cabin when we spotted a sheet web draped across an arching seed stalk of rye grass. We had by this time seen enough spider webs to last a lifetime, but this one was special, if only because we saw it in a different way. The criss-cross meshwork of fine strands was lined with tiny droplets of dew. Here and there at intersections of the web the drops had coalesced into larger spheres. The whole web was a creation of great beauty, and we felt lucky to have caught a glimpse of its ephemeral form. But at close range it was no longer a spider web covered with dew; it became a galaxy of celestial spheres, as surely as if we had seen it on a journey through space.

Why is a suspended drop of water spherical? Why are the planets spheres, and the stars? And the salmon eggs you have fished with from the shore of a lake, and many single-celled algae and protozoans and bacteria, and the bubbles you used to blow as a child? Why is the same drop of water no longer a sphere when it rests on a blade of grass? Why are we not shaped like spheres ourselves?

Once again we began reading, and once more we discovered that pathways from the meadow are many. A sphere is a remarkable thing with even more remarkable implications. Of all geometric forms, the sphere has the least surface area for its volume. Because of surface tension and cohesion of molecules one to another, any drop of liquid suspended in space assumes the shape of a sphere. Planets were once liquid, too, but in the case of such immense masses another force comes more strongly into play, the force of gravity. Many celestial bodies consist entirely of superheated matter in gaseous form, and it is the peculiar atomic force of gravity that holds them together and prevents the particles of matter from dispersing wholesale through space.

It has been known since the days of Galileo and Newton that it is the gravitational attraction of the larger mass of the earth for the smaller mass of a raindrop that causes rain to fall. It is this same force that holds captive the atmosphere and water vapor of the earth, without which life could, of course, not exist. And it is the opposing forces of gravity and inertia that control the orbits of the moon around the earth and the earth around the sun.

Many paths lead from a single meadow.

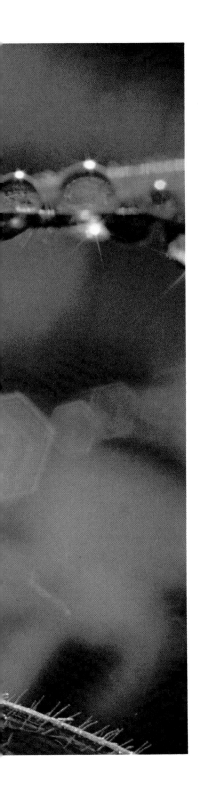

A tiny drop on a grass blade, we discovered, holds the story of surface tension, of water's movement through soil and roots to the twigs of spreading oaks, the story of blood flowing through our capillaries.

Thus, gravity is in no small way responsible for the rain that nurtures the meadow, and the seasons and the weather that create the rain.

It is gravity, also, that causes a sphere to flatten when it comes to rest on a solid surface. Perfect spheres are rare within our experience, and perhaps in the larger universe as well, because of other forces constantly at work on them. A raindrop is not spherical because of the friction of the air through which it must fall; a bubble is distorted by the tiniest breeze; the planets themselves are broader at the equator than on their polar axis because of forces created by spinning. Even the dewdrops on our spider web could not have been exactly spherical because gravity was pulling them, too, against the web strands which held them.

Living things have not evolved in a vacuum. The substance of life has been subjected throughout evolution to the same physical, chemical, and mathematical laws that control the form and behavior of matter and energy wherever they exist. Just as small masses of liquid are controlled more by principles of surface tension while larger masses are governed more by gravity, so it may be with living things. A single cell, or the egg of a fish, can be spherical, if no other paramount factors are at work, because it is small enough and fluid enough to be shaped by the forces of surface tension. But there is a size beyond which an organism enters another domain where gravity is the dominant force. The spicules of sponges, the exoskeletons of crabs and insects, and the bones of vertebrates answer, above all, to the force of gravity.

Relationships of mass have had a multitude of effects upon living things. They have deter-

mined that the mammoth reached about the largest size a mammal, with its particular body plan, can achieve on land, whereas in the sea, buoyed up by the greater density of water, whales have evolved which are larger even than the great aquatic reptiles of the past. Relative mass has in part determined that insects can grow no larger than several inches; that a flea can jump much farther, relative to its mass, than can a grasshopper, and a grasshopper farther than a man. These relationships, too, are responsible for the fact that smaller organisms can carry weights in greater proportion to their own than larger organisms can, that they can move their wings and legs far faster, and can make sounds of a much higher frequency.

Our quest of the spheres led us even farther into the biological implications of physics and mathematics. There exists a very simple principle that states that the surface area of a sphere increases as the square of the radius, whereas the volume increases as the cube of the radius. In simpler terms, as a sphere grows larger, its volume increases faster than does the area of its surface. Though animals may not be spheres, this principle of mass relative to surface applies nevertheless, so that if an elephant is fifty times the length of a mouse of similar design (and they are of similar design because they are, after all, both mammals), by simple mathematics (50^3) the elephant has a mass 125,000 times greater than that of the mouse.

It happens that both moisture and heat are lost from the body of an animal through its surface area. The greater that area, the greater the loss, which is in itself relative among all

Life is subject to laws of physics and mathematics.

animals, but it is the amount of loss per volume, or per mass, that is of critical importance. The smaller the animal, therefore, the greater is the potential loss of both heat and moisture.

The termites we saw in the ground and beneath rotting logs were soft and delicate. Like the majority of the organisms we found in the soil, they have no way of protecting themselves from the ever-present danger of water loss except to stay within the protection of their damp environment. But when we watched the kings and queens emerging from the ground for their nuptial flight, we were witnessing another adaptation, for their bodies were enclosed in a much sturdier, heavily pigmented exoskeleton, which enabled them, at least for the brief period of their exposure, to withstand light and the heat and dryness of the open air.

The harvester ants of the grasslands were active during the hottest period of summer, but they left the shelter of the ground only in the cool of the night, whereas the deer had no other shelter during hot days than the shade of an oak—and needed no other. So it was, too, that on the coldest night of the year, the meadow mice remained in their insulated burrows where their body heat was conserved, while the deer stayed out in the open with only their fur to protect them from the cold.

The greater discrepancy between mass and surface among the larger animals reduces the problems of heat and moisture loss while at the same time it raises some other difficulties that must somehow be surmounted. Many of the physical and chemical reactions that take place in living things are surface reactions. Food absorption and breathing, for example, must both take place through membranes that separate the outside of an animal from the inside.

For many small organisms, these processes take place directly through the outside skin. Numerous tiny aquatic insect larvae, especially those living in cold, oxygen-rich water, can absorb oxygen and give off carbon dioxide through their skin; even the breathing of frogs and salamanders is aided by this simple system.

Horsehair worms—strange, moving, wire-like strands frequently encountered in puddles or streams, creatures that legends claim are horsehairs come alive—are in reality internal roundworm parasites of grasshoppers, katydids, and crickets. For the duration of their sheltered lives they have no specialized structures for either breathing or feeding. Immersed as they are in the tissues of their hosts, they absorb what they need and give off their wastes directly through their body walls.

Single-celled plants and animals similarly need no highly specialized structures for most of their physiological processes because, like each cell in the body of a larger animal, all the necessary exchanges take place through the cell membrane itself.

As soon as an animal reaches proportions too large for simple diffusion to adequately distribute chemical intake and output, some other system must be devised. An animal of great mass has an enormous number of cells to service, each needing, by and large, the same materials for life that a protozoan needs. If all animals were simple spheres, as their mass and the collective needs of their cells increased, the ratio of surfaces available to meet these needs would decrease. What is obviously required, then, is a greater surface area than their mass would otherwise possess.

A large animal has an enormous number of cells to service.

141

*In a simple spider web covered with dew we
began to see a galaxy of celestial spheres.
Besides the shape of planets and protozoans, the
size of insects and whales, the behavior of mice
and termites, the structure of lungs and honeycombs,
how many facets of the universe have been
affected by the same laws of physics and mathematics
that control the properties of spheres?*

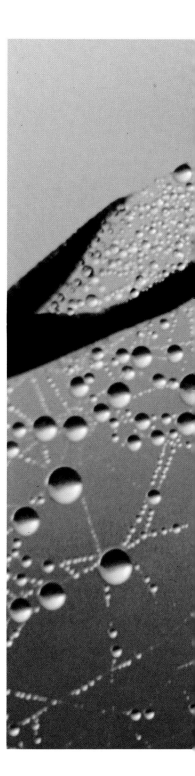

Increased surface area for breathing has been achieved through the evolution of gills and lungs which may contain, in the larger animals, an incredible number of surface convolutions. The same system of folded and elaborated membranes has been employed in digestive systems. Like a terry-cloth bath towel, our intestinal walls have thousands of protuberances, villi, which increase the absorptive surface.

These are, in a sense, outside surfaces which are usually folded inward in order to expedite the absorption of food or the transfer of gases. Once these substances have been taken within the body, however, they must reach every cell thoughout its volume, and so, very soon along the scale of size, animals have developed circulatory systems in order to aid diffusion and increase the absorptive area within the interior of the body itself.

Even a single cell has internal surfaces of various kinds, each generating its own kind and amount of surface energy. There seems to be some optimum balance between surface and mass in cells at which stability is achieved. The maintenance of this equilibrium apparently lies at the heart of the fact that cells, whether free-living or a part of a many-celled organism, do not vary much in size. In this respect, all living things are constructed from basically the same building blocks.

To use again the example of the elephant and the mouse, it has been found that the nerve cells of the elephant are only twice the diameter, and therefore eight times the mass, of those possessed by a mouse. Calculating from this, it is estimated that the elephant has 15,000 times as many nerve cells in its body, while the cells themselves have remained only slightly larger

Elephants and mice have cells of nearly the same size.

versions of the same units.

Our consideration of cells led us in still another direction in our investigation of spheres. Cells can be spherical when they exist singly, but they can no longer be spheres when packed together, or space would remain between them. If you took a series of circles of the same size and pressed them together on a single plane, they would become hexagons. A series of identical spheres, like balls of clay, similarly squeezed together, would become but three-dimensional versions of this geometric form.

The actual shape of cells in different tissues varies greatly, due to many additional factors, but it is this simple geometric consideration of closely packed spheres that nevertheless accounts for the very common and widely scattered occurrence of hexagons in the natural world. The combs I had seen in the fall, dug from the underground nest of the yellow jacket, and the storage cells of the honeybee both owe their hexagonal structure to this phenomenon. The insects begin the construction of their cells as cylinders, but it is thought that very soon they must respond to the equal pressures of others around them doing the same thing. Bees' wax and wasp paper are pliable materials, and they give under the equal distribution of tensions just as surely as soft cell walls do, or the walls of soap bubbles when they are forced together into the smallest amount of space. Whether a cause or an effect of this law, the fact is that the bees, in following the hexagonal pattern, use the least possible material to construct the necessary chambers.

By the same principle, the facets of the compound eyes of the bees and wasps themselves are also hexagonal. Simple variations upon this theme are responsible for the shape of spray-foam bubbles blown up on an ocean beach, the shape of spaces between the veins of an insect wing or a leaf, the configurations of many closely packed mineral crystals, and microscopic radiolarian skeletons constructed with extraordinary mathematical accuracy.

How many times in your life have you seen droplets of dew on a spider web, thousands of them almost identical in size strung evenly along the strands like strings of pearls? Have you ever asked yourself why the drops are so symmetrically arranged?

By the time Maggie and I finally asked the second question, we were almost embarrassed to answer the first, for we realized how easily we had taken this commonplace observation for granted. But once we began exploring the mathematics and physics of water drops, we were not long in arriving at both puzzle and its solution.

If you study an orb web carefully, you will see that dew adheres predominantly to the spiral catchlines and not to the spokes, the foundation lines, and other structural components of the web. This observation offers the first clue. If you were to examine, then, how the spider manufactures the sticky part of her web, you would find that as she ejects the silk and it quickly hardens, she also secretes another substance, the glue, which remains liquid for a long time in order to ensnare her prey. Here is our key to the physics of the process.

There exists in the science of physics a principle sometimes called the law of minimal areas. It is precisely by this law that water forms itself into spheres, because the sphere is the volume with the least amount of surface area. When the glue is secreted upon the web, it is spread evenly as a long, fluid cylinder around the strands. As this cylinder grows in length, it becomes unstable and tends to break up into a configuration of spheres which are stable. Since this happens in a regular fashion, as the length of the glue sheath approaches about three times its diameter, droplets of equal size and spacing form along the web.

Because of the viscosity of the glue, however, and its ability to adhere to the web, the sheath may not break up into droplets, and yet it might be selectively advantageous for it to do so. Spheres of glue, because of their smaller surface area, dry more slowly and are more effective in holding prey in the web. It is known that small vibrations will precipitate the sphere-forming process, and, for whatever reason, vibrations are exactly what the spider employs. Many species, as they complete each section of the line between spokes, are known to twang the web with one of their hind legs in order to precipitate and distribute the droplets of glue.

The dewdrops we see on a damp morning are merely larger representations of these tiny formations on the web. Just as raindrops tend to precipitate around tiny nuclei, such as dust particles suspended in the air, so also do the spheres of dew condense around the droplets of spider glue.

The same law of minimal areas, to some degree, also shapes the splash patterns of raindrops hitting the surface of a pond, and, by further extension, may even influence in some obscure way the shape of many protozoans.

Dew on a spider web reflects a principle of physics.

How many times have you seen droplets of dew on a spider web, thousands of them almost identical in size strung evenly along the strands like strings of pearls? Have you ever asked yourself why the drops are so symmetrically arranged?

We began to understand more fully than ever that a single drop of water, either directly or by implication, reflects many facets of the universe.

After these lessons among the orb webs, our attention was drawn elsewhere for a while; we were diverted from the subject of water drops by the many exciting activities of spring. Our spider-web investigations resumed quite unexpectedly, however, when I made a small discovery with my camera. The morning had been a damp one. I had been out early photographing and was returning home when a faint rainbow glint caught my eye among the strands of a web near the old road. I realized at once that I had noticed the colors in other webs before. Since I had never explored this phenomenon with my camera, I decided to try.

I located several webs in the vicinity which, when viewed at exactly the right angle, gave forth bright flashes of color. Maggie and I had many times seen dewdrops refracting sunlight, sparkling red or green or blue as we walked past them at just the right angle relative to the sun. But since these webs were now dry of dew, I could only guess that the much smaller beads of moist glue on the web strands were acting as minuscule prisms.

To my astonishment, as I probed within the meshwork of silk with my camera, I witnessed a glorious spectral display not otherwise visible. As the webs appeared more and more out of focus, each tiny spectrum was spread out until it became a fragment of the rainbow itself. The minute prisms along the strands were apparently situated at slightly different angles, for the resulting spectra represented various seg-

ments of the total range of colors. Lined up together, band after band, they created a dazzling spectacle indeed.

As we discussed my observations later, we realized that we knew very little about refraction of light and even less about the nature of light itself. We casually opened an old physics text gathering dust on our shelves and began, once again, a new and entirely unanticipated journey.

We found that Sir Isaac Newton was the first scientist who began experimenting with prisms and the refraction of light. Though he himself was not aware of it, Newton made some important observations which would later demonstrate that light travels in waves. In order to understand the behavior of light waves, we had to learn something about other kinds of waves—those in the ocean and smaller bodies of water, and those in the air. This investigation led us into the vast subject of sound. Little had we realized at the beginning that a spider web would raise questions about the song of a bird, the chirp of a cricket, or the elaborate vibrations of a symphony orchestra.

We learned that sunlight passing through a prism at an angle is separated into all the colors of the rainbow because each color represents a different wavelength of light. The red end of the spectrum consists of longer wavelengths, which are slowed, and therefore deflected, less in their travels through the prism than are the shorter wavelengths of blue and violet. The rain, with its millions of tiny units, acts as a giant composite prism, producing the familiar rainbow when the sun emerges from the last clouds of a storm.

But visible light, we discovered from our

A spider might inspire one to study Isaac Newton.

reading, is but one tiny portion of a much larger rainbow, a gigantic spectrum of other electromagnetic waves, which we can detect only through precise technology. The lengths of the waves we see as colors in a rainbow are extremely small. They range from .000028 to .000014 of an inch. Beyond the deep red of the visible spectrum lie heat waves, which reach a wave length of .01 of an inch. Radio waves have a length ranging from 600 to 1700 feet, while the longest electromagnetic waves range from 15 miles to hundreds of miles.

Below the violet end of the spectrum are ultraviolet waves as small as .0000005 of an inch. Smaller even than these are X rays, with a wave length as minute as .00000000038 of an inch; and gamma rays are shorter still.

The sun gives off a continuous stream of electromagnetic waves of many kinds, but heat and light are the only ones that we are biologically equipped to perceive. As flooded as our meadow could be with both on a hot summer day, the entire earth receives only one in two billion parts of what the sun sends forth in all directions through space.

Life has not developed an ability to handle excessive quantities of either ultraviolet radiation or the less understood cosmic radiation, because throughout most of its history, life has been shielded from these rays by the atmosphere. There is some conjecture, however, that in times past the amount of either or both of these forms of radiation that penetrated the atmosphere greatly increased for brief periods. Some scientists surmise that rather sudden changes in the composition of living things, such as the extinction of the great reptiles, may have been due to these shifts in radiation.

When ultraviolet rays strike the atmosphere, the widely scattered molecules of the thin, upper air are broken into charged ions. Through centuries of scientific experimentation, which began with Newton and his prism, we have learned to use the ionosphere as an enormous spherical mirror surrounding the earth, for it is from this atmospheric layer that all but the shortest radio waves are reflected back to earth.

The first attempt to measure the speed of light was made by Galileo over three centuries ago, in an experiment involving men with lanterns standing on distant hills. Little did he know that the lantern light traveled at a speed equivalent to eight times around the earth in a single second! It was not until 1676 that the Danish astronomer Roemer provided the first evidence that light had a finite speed, and the first laboratory measurement was not made until another two centuries had passed. The speed of light has finally been fixed at 186,000 miles per second.

By means of a long chain of experiments with light, astronomers have measured the diameter of distant stars, their atomic composition, their temperature, and the direction and speed of their motion through space. Studies with light have also revealed the answers to many mysteries of chemistry and the nature of the atomic origin of radiant energy, and have provided the source for Einstein's theory of relativity.

When Newton began experimenting with prisms for the first time, he initiated an avenue of study that has contributed more to man's understanding of the physical world than any other scientific pursuit, for light is the one

The study of light has led to man's understanding of atoms.

149

The spider, which knows the meaning of every web vibration, cannot perceive the beautiful spectra that illuminate its world. And yet man, whose understanding of light extends to the distant galaxies, will never know first-hand the feel of an insect tugging at the web-strands.

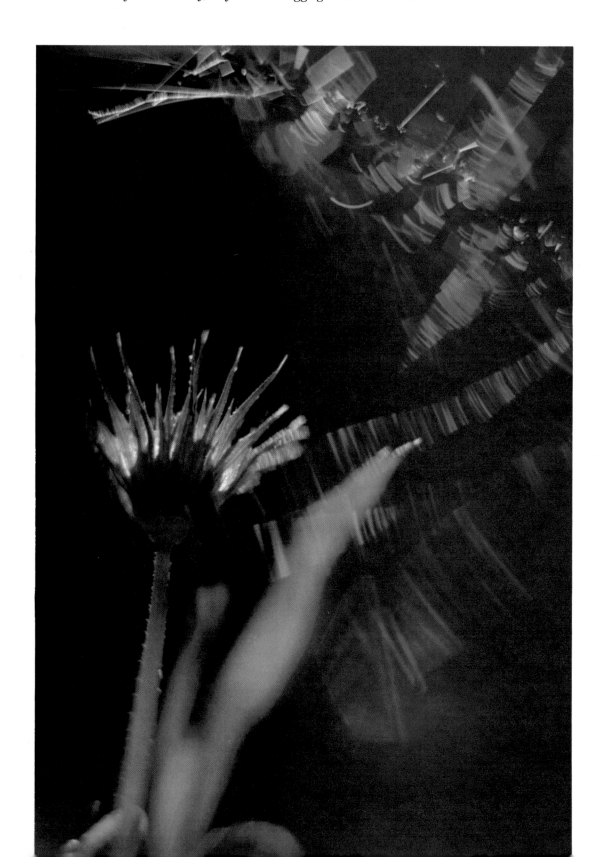

medium by which we can travel both into the atom and outward to the stars.

As Maggie and I continued our reading, we thus saw opening before us the awesome sciences of astronomy, chemistry, and nuclear physics, even the most cursory understanding of which would entail a vastly greater familiarity with the language of mathematics than either of us possess. Nevertheless, there they were, atoms and stars, two distant worlds we had touched in exploring the mysteries of a spider web.

M an has the ability not only to see rainbows in a spider web but also, by mental exploration, to follow that light to its source. And yet the spider, which is blind to these spectral displays and unequipped to perceive their significance, can feel the pull of an insect on its webstrands by means of sensitivities man does not possess.

Compared to most other species, especially the ancient spiders, we are newcomers on this planet. We have our own adaptations just as spiders have, but in spite of our unique abilities, we have no way of knowing which of us, spider or man, will survive in the long run. We have evolved the ability to learn and the capacity to create, but we may never learn to create that single element most important to survival, the future itself. Our vision of the future is shaped by our understanding of the past. It must be, for the natural laws that control all processes in the universe are, as far as we know, timeless.

As our third year progressed, Maggie and I became more aware of the importance of time.

Man may never learn to create the future.

We began to develop a sense of history. This interest in the past was not entirely intellectual in origin. In fact, like most notions, it began with an experience, the sort of happening that creates impressions greater in significance than the incident itself.

A jackrabbit had suddenly entered our lives. He was a week old. Friends had come upon the mother freshly killed by dogs. She was pregnant. In the hope of saving the young, they opened her body and found six babies; one was still alive. After a week of careful tending, the infant was given to us.

It seems to be the nature of man to name members of his family, even if they are temporary visitors from the wild. Under the circumstances, it seemed appropriate to call the rabbit Hercules. Though he was still very young, Hercules was fully clothed in fur and active. He ate well and grew quickly, both in size and strength.

Hercules lived with us for several months. He would romp for hours around the house

For how many millennia has the jackrabbit been evolving its long jumping legs, protective coloring, sensitive ears, and superb peripheral vision? And how many jackrabbits have been consumed by predators in the selective process by which these features have been refined?

A simple question about the fiddlehead
of a fern and the spiral of a snail led us to the
history of mathematics, art, agriculture—
to the history of civilization itself. We
found opening before us a vast web of knowledge
as complex and dynamic as the web of life.

Through what series of mutations and selective events have filaree seeds evolved the ability to plant themselves?

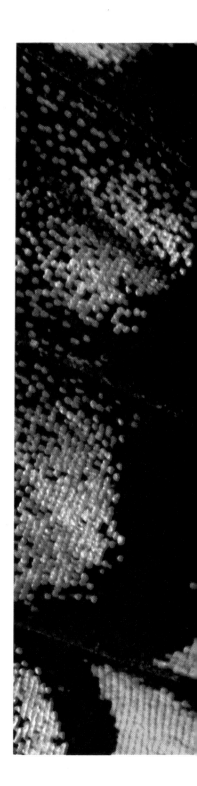

Every microscopic scale of a butterfly's wing
is precisely placed by genetic code—so precisely
that the pattern scarcely varies throughout
the entire species. What a history this
arrangement must contain, if only we knew how
to translate it into terms we could understand.

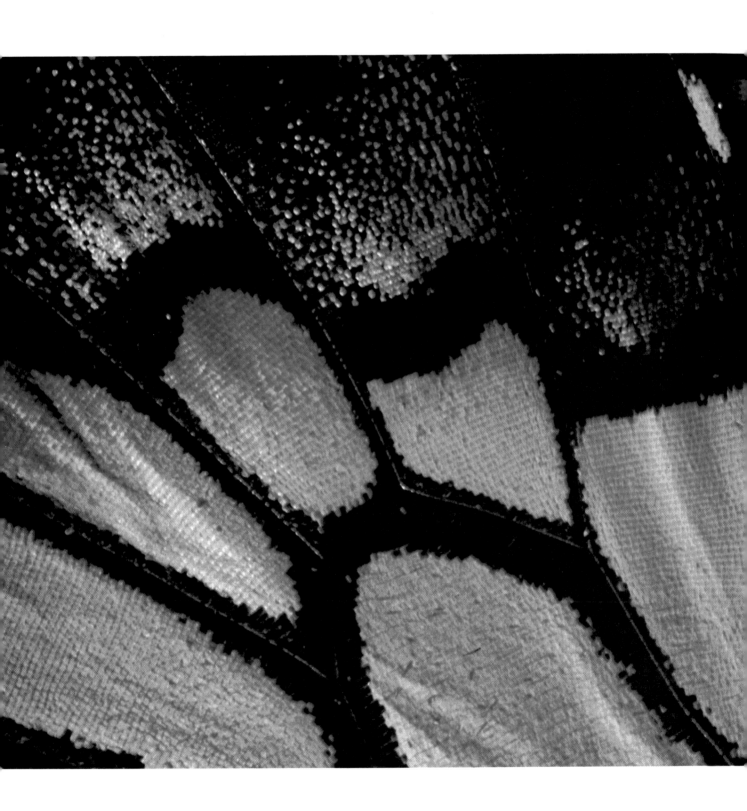

Behind the intricate structure of a feather, we may see the lizard's scale,
but concerning the evolutionary link between them, the fossil record is silent.

and welcomed a warm lap and gentle stroking. We became intimately familiar with him in a way that would not have been possible had he been separated from us by the distance of wildness. We began to see, however, that the closeness we shared was more an illusion than a reality. Hercules was not like a docile, domesticated cat or dog; he was a hare of ancient lineage whose ancestors had never known the taming hand of man.

We watched him explore the house. His long, sensitive ears were always up, scanning forward and backward and to the side; his nose and whiskers were constantly active. We saw that his large, keen eyes, set well back on the sides of his skull, gave him superb peripheral vision, and that his long hind legs were perfectly suited to swift movement and agility. Hercules would often "freeze" at the slightest new sound. Had he been out among the dry grasses, he would have been nearly invisible, so superbly would his brown coat have blended with his surroundings. His stillness was often prompted by stimuli we ourselves were unable to perceive. But if the stimulus was strong and sudden enough, like an abrupt movement or the slamming of a door, Hercules would leap straight up the wall and come down quivering, all senses alert and every nerve finely tuned.

Jackrabbits are herbivores. Like all rodents, they lie near the bottom of the food pyramid. For how many millennia have they been evolving protective coloration, sensitive ears, superb peripheral vision, long jumping legs, and a rapid rate of reproduction? How many jackrabbits have been consumed by predators in the selective process by which these remarkable features have been refined?

For the first time we were becoming acutely aware of the immensity of time within which evolution operates. Hercules was unmistakably wild. No amount of taming within his lifetime could prevent his instinctive reaction to the slamming of a door. We were pitting a few months against millions of years. Hercules, we knew, must return to the grassland.

We took him to a wild hillside and quietly climbed the grassy slope. At first he remained in Maggie's lap after she released her hold. Tentatively, quivering with fear of the unknown, he began to explore. Gradually gaining confidence, he ventured farther and farther from us until he disappeared over the ridge. He never looked back.

It was nearly dusk when Hercules left us. As we returned home, we remembered how a jackrabbit had awakened us from our preoccupation with the minute during our first year of exploration in the grasslands. Once again a jackrabbit had aroused in us a new awareness. We began reflecting upon the diverse forms of life with which we had become acquainted. They seem to raise an endless stream of fascinating evolutionary questions.

We had seen, besides the jackrabbit, a host of other creatures that exhibit protective coloration. Wolf spiders, caterpillars, meadow mice, gopher snakes, fence lizards, and many kinds of grasshoppers are superbly camouflaged. On the other hand, we had learned that monarch butterflies and ladybird beetles are distasteful to most of their potential predators and advertise this quality with brilliant orange coloration. We had seen a skunk, too, which treats its

Living things gradually change through mutation and natural selection.

161

Living things have developed different responses
to the rhythm of night and day. The morning sun
awakens the fly while it sends the slug in
search of a damp place to hide.

enemies to a disagreeable surprise and makes its presence known to them thereafter by means of the nocturnal warning colors, black and white.

How did these very different protective devices evolve? What evolutionary history lies behind the fact that tarweeds are unpalatable, that thistle seeds have parachutes, that mistletoe is a parasite, and that termites have castes?

Gradual changes in the anatomy, physiology, and behavior of living things are primarily the result of two factors: mutation and natural selection. Mutations are changes in the genes themselves, which in turn affect some characteristic of the individual. An organism is so delicately balanced that even the most minute change in its makeup can cause death. Most mutations are, therefore, lethal, but once in a while an individual survives some small alteration in its design, its body chemistry, or its behavior. The selective factors at work in the environment ultimately determine whether this individual is more or less suited for survival than others of its kind, and therefore determine whether the mutation will persist and spread through subsequent generations or be eliminated along the way.

The rate of evolutionary change depends largely upon the frequency with which mutations occur and the nature of environmental factors to which any given species is subjected. Though the mutation rate varies from species to species and environments vary from place to place, the process is most often incredibly slow, building change upon minute change over hundreds of thousands of years.

Think of all the different kinds of living things you have seen in your lifetime. There are over a million species of insects alone on earth. Imagine the billions upon billions of life forms that have existed in the past, constantly changing, dying out, or persisting to become eventually the species of today.

We can read the trends of the past in the fossil record and in relationships that exist in the structure of things presently living. But the minute genetic changes and selective events from which that history has been built have left no record of their occurrence. Had a tally been kept, their number would completely surpass our comprehension.

Each time we walked over the fields and hills, we found in some familiar plant or animal a new wonder of evolution. We studied the seeds of filaree, a wild relative of the garden geranium. Each flower produces five seeds attached to filaments that are joined together in a long, pointed fruit. As the filaments dry, they begin to curl, seed end first, and the five form a beautiful pinwheel balanced atop a slender central column. They finally wind all the way off and fall to the ground.

Each filaree seed is sharp and barbed, superbly adapted for dispersal in the fur of an animal or the sock of a hiker. What is more remarkable, the spiraled filament uncoils when exposed to moisture and winds up again as it dries. During the course of this motion, the arm of the coil levers against the ground so that the seed itself is turned. With the alternation of moist nights and dry days, the seed gradually moves across the ground, finds some crack, and works its way into the soil. Thus the seed of filaree actually plants itself. Though it hardly seems believable, this ability evolved entirely

Evolution builds change upon minute change.

through the chance occurrence of mutation and selection.

During the summer we collected a few butterflies and studied their intricate wing patterns. Many were familiar to me. As a child I raised most of the local species from caterpillars. I have felt ever since that no one's life could be quite complete without having watched, at least once, the transformation of caterpillar to chrysalis and chrysalis to butterfly. As the delicate adult emerges it pumps fluid through the veins of its dwarfed, compressed wings. The whole pattern is there, every microscopic scale of it; and as the wings unfurl, the membranes stretch, carrying the colorful little flecks with them. The scales take their final place, the wings broaden, the design is complete. Once the wings have dried, fluids cease to flow in their veins, for the wings are not alive and neither are the colorful scales that adorn them.

Every scale of a butterfly's wing is so precisely placed by genetic code that the larger pattern scarcely varies throughout the entire species. That these beautiful designs emerge from the reorganization of a caterpillar's tissues will always remain one of the miracles of evolution.

Late in the summer we noticed an abundance of feathers scattered among the dry grass stalks. Some we recognized: quail, mourning dove, sparrow hawk, owl. Most were unidentifiable, but all were immensely beautiful. They quivered in the slightest breeze. Once in a while one would leave its grass perch and drift away across the meadow. As the afternoon sun illuminated them, they seemed, in the squint of an eye, to be hundreds of small sailboats dotting an expansive sea.

It was time for the summer molt. The weather was hot and dry, a perfect time for the birds to grow new feathers because the old ones would not be sorely missed. We found long, sturdy flight feathers, curving body feathers, and soft, puffy down feathers. The functions of flight, insulation, protective coloration, species recognition, and courtship display are all served by feathers of various designs. Each is structured to best fulfill its particular role—another remarkable result of the evolutionary process.

Birds apparently evolved from reptiles. The oldest known fossil of a bird is *Archaeopteryx*, which lived one hundred fifty million years ago. This strange creature had teeth, claws on its wings, and a long tail. In most ways *Archaeopteryx* was very much a lizard, but curiously it had feathers almost exactly like those birds possess today.

Like the wings of a butterfly, feathers are not alive. They grow from the skin like the hair of a mammal or the scales of a lizard. Once they are formed, they become dead tissue. Like a mammal's hair or a lizard's outer skin, feathers can be repaired only by being replaced entirely.

Since birds seem to have developed from reptilian ancestors, and since feathers, like scales, are products of the skin, we might assume that feathers evolved from reptile scales. Their actual origin remains unknown, for the fossil record is silent. *Archaeopteryx*, with its fully developed feathers, came too late in the evolution of birds to offer a clue.

Turpentine weed provided us with still another amazing example of the refinements the evolutionary process has produced. Like

Like butterfly wings, feathers are not alive.

Grasses bloom in the spring, soaproot late in
summer; bumblebees visit flowers by day, moths
by night; and turpentine weed mechanically
brushes its insect visitors with pollen.
A mystery of evolution exists at every turn!

tarweed, this peculiar plant grows during the hottest months of summer, when nearly everything else is dead or dormant. It, too, is sticky with glandular secretions, and it gives off the pungent odor of turpentine. The deer and rabbits seem to leave it alone entirely.

For some reason we had missed turpentine weed during the previous years, perhaps because it is not common in our area. We found the first plant in full bloom not far from our cabin. It was late October. The first rains had already fallen, the nights were once again cool and moist, and the little aromatic plant was covered with dew. It was the nature of the flower itself, however, that really caught our attention.

A number of blossoms were open on the stalk. They were delicate lavender, about half an inch long, shaped roughly like the flowers of mints or snapdragons. From each blossom, the pistil and stamens protruded in a long, curving arch, a peculiar feature unfamiliar to both of us.

Before we had time to unravel the mystery, a bee arrived at the plant and began visiting the flowers one by one. To our astonishment, as the bee settled on the "landing platform" and entered the throat of the blossom, the stamens and pistil swung down and brushed the bee's back. The flower is so structured that the bee mechanically levers the pistil and stamens as it enters the flower. Furthermore, the pistil extends beyond the stamens so that pollenation is more often effected by an incoming than an outgoing bee. In this way not only is pollenation assured, but also the chance for cross-fertilization is increased.

What length of time has been required for the refinement of this remarkable mechanism

Specialized flowers have been a long time in coming.

for pollenation? Flowering plants first made their appearance around one hundred forty million years ago. The earliest flowers were simple in design, rather like a magnolia or a buttercup. When you compare a buttercup with turpentine weed, it is not hard to imagine that this highly specialized flower has been a long time in coming.

As we explored our familiar meadows with evolution in mind, we became aware of behavior as well as structure. We watched the courtship of jumping spiders—an elaborate process, different for each species, attended by much leg-waving and dancing. As we read more about these fascinating little creatures, we learned that the retinae of their two largest eyes are capable of rotation and that the angle of rotation almost exactly corresponds to the angle of the legs during courtship. In other words, anatomy and behavior are so superbly allied that by means of the geometry of the courtship dance and the corresponding structure of the eyes, the female is able to distinguish the advances of her mate from the movements of potential prey.

The rhythms of night and day, summer and winter, rainfall and drought have exerted their own selective forces upon living things. Flies, dormant during the cool of the night, wake up at sunrise, just when snails and slugs, which cannot tolerate the heat of the day, are seeking a damp hiding place among the grass clumps. When the annual grasses and wildflowers are long since dead and brown, the delicate shoots of Indian soaproot emerge from bulbs deep in the dry, cracked earth. In the evening, as bees and butterflies search for places to spend the night, the delicate lily flowers of the soaproot open, their white petals advertising for a different set of pollenators, the moths of the night. And while summer heat brings drought and dormancy to the grasslands of California, it brings rain and growth to the grasslands of the high mountains and the eastern states.

The more grassland organisms we became acquainted with, the better able we were to see evolutionary themes, the larger relationships which exist among living things. We gradually became familiar enough with the system of classification to begin to recognize many orders and families of insects and families and genera of plants by their anatomical similarities and differences.

By and large, classification is an arbitrary human device for rendering a vast variety of forms comprehensible and manageable. It is an aid that expedites our own communication. In the case of living things, however, the system represents also an attempt to interpret the actual relationships that exist between one form and another. In other words, it is an effort to represent, as accurately as possible with the fragmentary knowledge at hand, the history of evolution.

In reality, there are no phyla, classes, orders, families, or genera in nature. The only real category of living things is the species itself, a population of individuals having basically the same characteristics and capable of interbreeding. These individuals are acted upon by natural selection; they alone evolve. A species does not suddenly produce or become another species; it changes to something new only as the individuals themselves gradually change. All the

higher categories in the classification simply reflect the pathways through which these alterations have occurred in the past.

Some similarities in the structure or behavior of organisms are due not to genetic relationships but instead to common factors in the environment. Whales, sea lions, sharks, and fish all have flippers or fins and streamlined bodies because of the common medium to which they have adapted, even though they are not at all closely related. Many desert animals are active at night because of the intense daytime heat, even though one may be a beetle, one a snake, another a mouse. Tarweed and turpentine weed both produce unpalatable secretions to ward off the summer-hungry deer, even though they are members of unrelated plant families. This kind of similarity is known as evolutionary convergence.

We had learned in our investigation of the properties of spheres that fundamental laws of the universe underlie the process of evolution itself. It is a result of physical law that the hexagon appears in the honeycomb of the bee, the compound eyes of an insect, the cell wall of a diatom. It is also a result of physical law that whales, sea lions, sharks, and fish are streamlined. What, then, can be said about spirals?

We had been noticing spiral forms throughout the year—the unfolding fiddleheads of ferns, the arrangement of disk flowers in the centers of wild sunflowers, the coiled shells of snails. Has the spiral form evolved in these unrelated organisms in response to some common law?

Our investigation this time took us into the

By and large, classification is an arbitrary human device.

169

The grasslands are a forgotten landscape,
recipient of man's worst abuses. They have
been overgrazed and broken by the plow; native
grasses have been replaced by wild oats and
weeds; wildflowers have been severely reduced
in number. It is not a pleasant history.

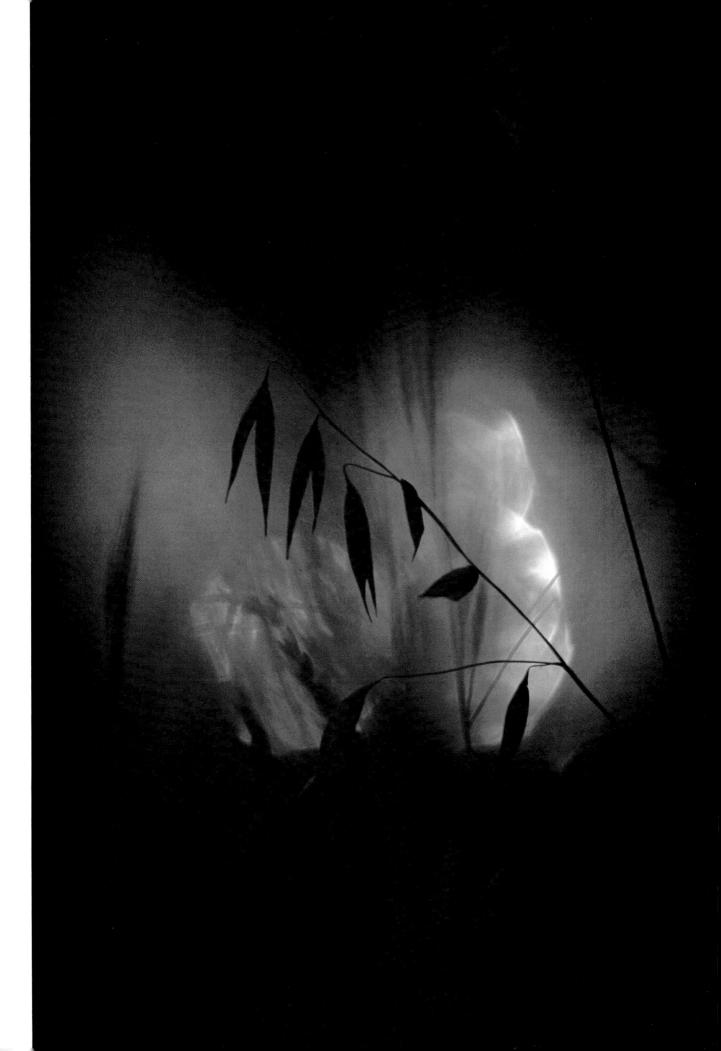

realm of mathematics. There, among the ratios and logarithms, we learned some surprising things about growth and form.

Draw a rectangle so that its sides are in the ratio of .618 to 1. You will find that by drawing a line through it, you can subtract a perfect square and have left a smaller rectangle of exactly the same proportions as the one you started with. It is possible to do this only with a rectangle of these exact proportions.

Continue this process systematically, marking a square within each subsequent rectangle until the space becomes too small to draw in, and then swing a curve through all the squares, from the inside out. You will have drawn a logarithmic spiral. This spiral is precisely the shape of a chambered nautilus. It is also the spiral that is expressed in the snail, the fiddlehead, the sunflower, and an immense variety of other living things.

Before the year 1220, an Italian mathematician named Fibonacci happened upon a series of numbers the implications of which may not yet be fully known. The series is derived as follows: $1+1=2$, $1+2=3$, $2+3=5$, $3+5=8$, and proceeds to 13, 21, 34, 55, 89, 144, etc.

If the terms of this series are set up as fractions, $1/1$, $1/2$, $2/3$, then placed alternately in two columns and the results expressed as decimals, we find that the results in one column get progressively smaller whereas the results in the other column get larger. In other words, they tend toward some common number. That figure, it turns out, is .618, the same number represented in the magic rectangle!

If you count the number of clockwise and counterclockwise spirals in the head of a sunflower, you will find that the results are suc-cessive numbers in the Fibonacci series. Count the leaves on a stem from bottom to top; when you come to a leaf directly in line above the one you started with, you will have arrived at one of Fibonacci's numbers. The series also expresses itself in the spiral design of pinecones, artichokes, and other botanical forms.

What all this means is still not clearly understood, but there is at least one feature common to all the natural forms whose design is based upon the logarithmic spiral: they grow from one end only.

Nautilus and snail shells, for instance, are added to on the outward edge. The shells grow in size but not in shape, as a cone would if it were added to in the same fashion. In fact, a snail shell is little more than a cone rolled up or coiled upon a vertical axis. The horns of a ram, the tusks of an elephant, the front teeth of a rodent, the claws of a cat, the beak sheath of a hawk, like the shell of a snail, all consist of dead material formed by living cells. They, too, are added to gradually from one end, and they all reflect in their form the logarithmic spiral.

Growing stems, pinecones, and sunflowers are not dead tissue. They are very much alive, but they, too, grow from one end only, from an apical meristem, or bud. As each new leaf, cone scale, or floret is produced, it becomes just as much an element of the past as each subsequent growth ring of the snail shell.

The logarithmic spiral and the Fibonacci series seem, then, to express mathematically a sort of dynamic symmetry derived from one-directional growth and the sequence of time this kind of growth entails. Or perhaps it is more correct to suppose that growth itself is an expression of universal laws of mathematics.

The ratio .618 to 1 is hidden in many natural forms.

As we studied the story of Fibonacci and his numbers, history took a firmer hold upon us. We both had musty, rather unhappy school-day memories about the pyramids of Egypt and various exploits of the ancient Greeks, but now, because of a simple discovery in the grassland, we found ourselves facing those very same gates of history with a feeling of exhilaration.

We learned that nearly five thousand years ago the Egyptians were developing an elaborate system of geometry from the simple mathematics they had been using to relocate farm boundaries after the annual floods of the Nile. Sooner or later this geometry became known to the artisans, and a primitive form of dynamic symmetry began appearing in works of art.

Several centuries before Christ, the mathematical knowledge of the Egyptians found its way to Greece. The Greeks began experimenting with the Golden Section, a line divided in such a way that the shorter part has the same ratio to the long part as the long part has to the whole line. This ratio is, once again, .618.

Euclid, Pythagoras, and other thinkers of the times were developing theories upon which much of modern mathematics would be based, but it was in art and architecture that principles of dynamic symmetry were most fully expressed and refined. The Golden Section was employed in thousands of statues and architectural structures, and rectangles close to the proportions .618 to 1 were used to an incredible degree of perfection in placing salient figures on pottery.

At the end of the Greek classic period, this superb form of design fell into disuse and was forgotten. A few centuries passed, and Fibonacci stumbled close to it, but it was not until about 1900—over two thousand years later—that a man named Hambidge at Yale rediscovered the principle upon which much of Greek design was based.

Imagine the excitement Hambidge must have experienced! The thrill of discovery is the same no matter what the discovery itself may be. Finding a living species new to science is no different from finding a new species in the fossil record. Discovering an idea new to man's experience is no more exciting, and may not be any more important, than rediscovering an old idea that has long been forgotten.

We experienced this exhilarating feeling that also accompanies the learning process, for although we had made no unique discoveries, we had found a few of the connections that exist between widely divergent bodies of knowledge. The simple observation of a snail in the meadow and a fern frond uncoiling had led us to the threshold of history. Through this single avenue we had approached the entire history of art, the history of mathematics, and, had we pursued the matter beyond the Nile Valley farmers, even the history of agriculture.

We began thinking about other things we had seen in the grassland. Where might they lead us? What about the soaproot we had seen blooming on summer evenings? Its bulbs were used in various ways by the local Miwoks and other Indian tribes, a simple fact that opens the door upon the entire history of the New World's indigenous peoples. What about the skunk, the ladybird beetle, or the tarweed? A study of their repellent properties would lead directly to the immense fields of chemistry and physiology, sciences that occupy their own places in human history. And what can be said

Whatever the discovery, the thrill of discovering it is the same.

about the grasses themselves, the fields upon fields of them we had walked and crawled through for three years? They, too, would direct us into our past because, in the form of rice, wheat, corn, or some other grain, grasses have provided the foundation for every major civilization the world has ever known.

All plants and animals known to man offer potential access not only to the evolutionary past but also to the recorded history of our own species, if for no other reason than because each has been collected by someone, named by someone, and may have been studied by many others during the course of time.

In browsing through the scientific literature, we found the names Wyeth, Godet, and Eschscholtz, preserved in *Wyethia* (mule-ears),

Man, like any other species, has the capacity to fail. But man also has some special tools— the ability to learn and the capacity to love.

Godetia (farewell-to-spring), and *Eschscholtzia* (the California poppy). We found also the names of Nuttal, Menzies, and Douglas, who described many of the plants of the West. In turn, a host of plants, including numerous species these men discovered and collected for the first time, were named by other botanists in their honor—plants such as the sego lily (*Calochortus nuttallii*), baby blue-eyes (*Nemophila menziesii*), and the blue oak (*Quercus douglasii*). Over and over again we found the name of Linnaeus, father of our system of classification and nomenclature, who named, besides the monarch, the painted lady, and the mountain lion, probably more species of plants and animals, and certainly a greater variety, than any other man. Who were these men? What has been written about their studies, their adventures, and the times in which they lived that wouldn't present a thrilling path of discovery to follow?

Scientific names contain more, however, than just an indication of the people connected with them and the events that circumscribed their lives. Nearly the entire biological nomenclature is derived from Greek and Latin and therefore constitutes a superb access to the history of language.

The genus *Bittacus* is derived from a Greek word meaning "beak," referring to the peculiar mouth parts of the scorpion fly we had seen in the foothill grasslands. A related Greek word, *psittakos*, means "parrot," from which psittacosis has been derived, the name of a disease which afflicts these birds.

The harvestmen we had watched among the grass clumps, are classified Philangida, from *phalangion*, Greek for spider. These words, in

Scientific names contain a wealth of history.

turn, are derived from *phalanges,* meaning lines of battle, ranks of soldiers, or, because of their linear arrangement, the bones of fingers and toes. There is even an Australian mammal, the phalanger, which bears this same root in its name because of peculiar second and third phalanges in its hind feet.

Nemophila, meaning to love the woods, not only perfectly describes the preferences of baby blue-eyes and other members of this genus, but also leads to a wealth of linguistic relationships. The word *phileo* appears in *philosophy* (love of wisdom) and *philology* (love of words). From *logos,* Greek for "word," and its Latin derivatives, we have received such terms as *analogy, homology, biology, zoology, entomology,* and many other *-ologies* describing areas of study and discourse.

Some biological names are fanciful or incorporate interesting bits of historical lore. *Ranunculus,* the buttercup genus, comes directly from the Latin and means "little frog," referring to the moist habitat preferred by most buttercup species. *Lupinus,* the lupine genus, comes from *lupus,* Latin for wolf, because it was thought that these plants robbed the soil of fertility. Concerning the behavior of both wolves and lupines, this notion is a myth, but it is part of recorded history nonetheless.

History, above all other disciplines of knowledge, offers us a sense of perspective. It gives us a certain distance from which we can see that ideas, like species, are not static but exist also within the dynamic process of evolution. A friend of ours recently called our attention to a passage he had found in an 1875

History offers a sense of perspective.

publication, *Plato, and the Other Companions of Sokrates.* Reading these words revealed to us how dramatically man's views have changed concerning the processes of life.

"The more brutal land animals proceeded from men totally destitute of philosophy, who neither looked up to the heavens nor cared for celestial objects: from men making no use whatever of the rotations of their encephalic soul, but following exclusively the guidance of the lower soul in the trunk. Through such tastes and occupations, both their heads and their anterior limbs became dragged down to the earth by the force of affinity. These men thus degenerated into quadrupeds and multipeds: the Gods furnishing a greater number of feet in proportion to the stupidity of each, in order that its approximations to earth might be multiplied. To some of the more stupid, however, the Gods gave no feet nor limbs at all; constraining them to drag the whole length of their bodies along the ground, and to become reptiles."

Several years ago I wandered into a damp, dimly lit used-book store in Singapore. I was rummaging idly when suddenly I caught sight of an ancient volume on a high shelf. Its pages were tarnished on the edges but its hand-painted illustrations were still fresh and beautiful. It was Volume V of *The Naturalist's Library, Foreign Butterflies,* published in London in 1837.

Recently Maggie and I read the introduction together, concerning the life and work of Lamarck, who had died not long before this volume appeared. Like many other men of his time, Lamarck was beginning to doubt the prevailing belief that the universe was static. In the process of groping for new explanations of the phenomena he saw at work in the world around him, he expressed a number of startling ideas, perhaps the most famous of which pertains to the evolution of living things.

"In like manner," Lamarck was quoted, "it is the desire and the attempt to swim, that had, in time, the effect of extending the skin that unites the toes of many aquatic birds, and thus the web-foot of the gull and duck were at last produced. The necessity of wading in search of food, accompanied with the desire to keep their bodies from coming in contact with the water, has lengthened to these present dimensions, the legs of the grillae or wading-birds; while the desire of flying has converted the arms of all birds into wings, and their hairs and scales into feathers."

From our present point of reference, we look upon Lamarck's theory as quite outdated. Though he has been proved right in his belief that species gradually change over long periods of time, the mechanism of change he espoused has long since been refuted. The author of this little volume blasted Lamarck's views as "licentious speculations" which "are of a highly fanciful description, and some of them greatly to be deprecated on account of their hurtful tendency . . . ," not because Lamarck's method of change seemed in error, but because he dared to question the immutability of species.

Just a few months before Volume V was published, Darwin returned to England aboard the *Beagle* with the ideas brewing in his mind which, when they appeared in public twenty years later, shook the religious and scientific communities to their very foundations.

Ideas are like genes themselves. They are constantly being altered and recombined, and they find acceptance only when the social climate is right. A great idea, like a new species, is not the result of a single change in thinking, but instead the product of many small mutations that were the work of countless men, most of them forgotten. They are a sort of culmination that survives only as similar collective changes in the cultural environment allow it to survive.

Unlike biological species, however, the few great ideas that have lasted throughout human history will change no further if, in fact, they correctly represent the laws of the universe. Taken together, these laws constitute a body of universal truth, which is, as far as we know, immutable. And yet, can we be sure? A hundred years from now will we discover that even the laws of the universe change?

As our third year drew to a close, Maggie and I revisited many of the places we had discovered and enjoyed. We went again to the foothills, basked in the sun of Indian summer, and absorbed the sounds of crickets chirping in the night. We watched thistledown blowing and young spiders ballooning; we heard again the pounding of the acorn woodpecker and the lonely cry of the coyote.

We visited our meadows near the coast and walked once again the same creek banks, hillsides, and deserted old roads. We saw yellow jackets flying in and out of their nests in the ground, old discarded bird feathers in the grass, fallen wild grape leaves turning scarlet, tarweeds, and the first autumn asters. In the evenings we listened to owls hooting from the

Ideas, like genes, are constantly being altered and recombined.

Only man can see the laws of the universe beyond the setting sun. He alone can
stare into a spider web and see, through its design, the web of life and the web of knowledge.
No creature on earth but man can examine his own soul.

limbs of the ancient oaks and watched deer browsing on the slopes against an orange sky. Each plant and animal was like an old friend; in seeing them again, we relived the joy we had experienced upon our first meeting. The year was ending, but not the search, for we knew that there would always be more to explore and more to discover.

Beauty, significance, and mystery surrounded us on every side, and yet, as had been the case for three years, we were almost always alone in our explorations. The grassland is a forgotten landscape, recipient of man's worst abuses and precious little of his interest and concern.

We learned that during the early days of settlement the grasslands of California were severely overgrazed by sheep and cattle. Ranchers at that time raised cattle for hides; they squeezed as many as possible on the rangelands because well-fed animals were not required in the production of skins. The sturdy, perennial bunch grasses that were native to our grasslands were largely eliminated; top soil was washed away, choking the rivers and filling the bays. Vast fields of spring wildflowers—some of the finest floral displays in the world—gradually dwindled and all but disappeared.

In place of the native plants came many kinds of weeds and annual grasses from Europe, including wild oats, which now dominate the grasslands of California. These species have shallower roots than the native perennials; they do not hold the soil as well, and they provide inferior fodder for grazing animals. Today, out of the hundreds of thousands of grassland acres in California, only a few remain unaltered by the hand of man.

Nor was the rest of the nation spared. Mountain meadows and desert grasslands alike were opened up for grazing. And where stock didn't tread, the plow did. It is not a pleasant history, least of all in the great prairie of the Midwest, where the ancient soil was broken by the plow and blown away by the wind, where man learned one of his hardest lessons.

Man, like any other species, has the capacity to fail. It is inherent in the selective process that not all notions succeed. Think of all the ideas, all the experiments throughout history that, like the giant reptiles of the past and countless other forms of life, have failed the test of time.

But man has a special tool, the ability to learn. He is the only creature on earth who can watch a deer browsing at sunset and see behind its structure the entire history of evolution. He alone can see the laws of the universe beyond the setting sun. Only man can stare into a spider web and see, through its design, the web of life and the web of knowledge. No other creature but man can examine his own soul.

Man also has the capacity to love. That which man loves, he tends to preserve. We live in a varied and beautiful world. To the extent that we become aware, with all our senses, of the diversity of forms and processes that surround us, we will find within ourselves the desire to preserve what remains of the earth's wildness.

We have reached a crossroads in the history of our species. Our numbers, and therefore our impact upon the planet, have increased, but so also have our knowledge and understanding. We must heed these things we have learned. This living earth is our only home, and our capacity to learn and to love is the only means we have to save it from ourselves.

*Nothing in the universe exists alone. Every drop
of water, every human being, all creatures in the
web of life and all ideas in the web of knowledge,
are part of an immense, evolving, dynamic whole as
old—and as young—as the universe itself. To learn
this is to discover the meaning of joy.*

PHOTOGRAPHIC NOTES

Photography is a tool which sharpens awareness and increases one's perception of the significant. It offers a means of selection and interpretation. In short, photography helps one learn to see, but what is even more important, taking pictures puts one out in the midst of things where experiences are direct and firsthand.

The photographic process itself is one of discovery from beginning to end. Seeing the negative just out of the developer or a box of slides just back from processing imparts its own feeling of adventure. One of the most fascinating and delightful experiences I had during our three years of discovery in the grasslands was watching my wife Maggie explore with a camera for the first time. The thrill of seeing with a new sense of focus, the anticipation of realizing both success and failure in her results, and the excitement of learning were often expressed in her face, and therefore in my memory. As a result of her efforts, two of Maggie's pictures appear in this book.

I have also found, in teaching nature photography, that no two people see things alike. When fifteen or twenty of us photograph for a day in the same meadow and later share our results, we are always astonished to discover in the pictures as many interesting and diverse styles of expression as there are people in the class. Yet the kind of equipment used seems to contribute to these differences only in minor ways.

My photographs, like anyone else's, represent my own personal interpretation of the natural world. They are as true to the substance and spirit of each recorded moment as I know how to make them. But they are no substitute for the real thing. They cannot replace the feeling of lying in a wet meadow, the sensation of numbness on a frosty morning, the smell of the first rain on dry grass, the call of the owl, the croak of the frog, the exhilaration and joy that accompanied the photographic excursions. To have the full experience, you must go there yourself.

I am probably more impatient with the mechanical aspects of photography than Maggie was during her first tries with the camera. Since I first began seriously taking pictures about eight years ago, I have resisted the equipment itself. I almost never use a tripod, preferring a rock or a log to support the camera for long exposures. All the pictures in *This Living Earth* were taken with one camera, a 35mm Nikkormat, and most with one lens, a 55mm Micro-Nikkor, sometimes with an M-ring or extension tubes added for especially tight close-ups. The only other lens I use regularly is a Soligor 135mm telephoto for long shots and, with extension tubes, for an occasional close-up.

All the equipment I use during a day's walk fits into a small canvas belt bag strapped around my waist. When I'm out exploring, I want to be mobile, unhindered by excess baggage. The beauty of a 35mm camera is its light weight and maneuverability. It places a smaller mechanical barrier between me and what I see than cameras of larger format would, and therefore interferes less with the total outdoor experience. In addition, no other camera system can compare with a 35mm single-lens reflex in terms of the ease with which it can be used for close-up photography.

One particular philosophy very strongly influences my approach to nature photography. I believe in the use of sunlight and all the magnificent moods that it creates. The colors of flowers and insects, of frost and dew that appear in this book may not be true to nature in the documentary sense, but they are true to the quality of the light and the mood of the moment, at least as far as the film was able to record them. With the exception of two photographs taken through a neutral density filter to reduce the sun's intensity, all the pictures were made with unfiltered, unaltered natural light.

Two phenomena of close-up nature photography that appear often in this collection of pictures raise questions so frequently that they deserve special mention. One pertains to the background hexagons, and the other to images of the sun itself.

Hexagons in the background of many close-ups are a function of optics and lens mechanics. Any highlight, such as reflections from dewdrops or shiny leaves, expands in size (and decreases in intensity) the farther away it is placed from the plane of focus. Out of focus, a spot of light assumes the shape of the iris diaphragm of the lens. My equipment makes hexagons if the diaphragm is closed down at all and circles if the lens is wide open.

Exactly the same phenomenon occurs when the sun appears in close-up photographs, because the sun is very much beyond the plane of focus. Since the real sun never appears in the shape of a hexagon, any close-up that includes it as part of the background must be shot with the lens wide open.

Under any circumstances, the full sun on a clear day is too intense to be included directly in close-up photographs. Therefore, all the close-ups of this nature in the book include only part of the sun, or, more accurately, part of the sun's light. They were all taken either at sunrise or sunset, just as the sun meets the horizon and is partially screened by the horizon itself or some distant object.

The denser the atmospheric dust and other impurities, the redder and less intense the sun will be at these moments and the less screening it requires for direct photographing. If the air is relatively clean, or if the sun is high already when it breaks over, say, a hill, then the light appears pale yellow or white and is so bright that most of it must be blocked out by some distant object. The size the sun assumes in the final picture depends upon how close one is photographing.

The more extension tubes or bellows one has behind the lens the larger the sun will appear.

Kodachrome and Ektachrome films with ASA ratings of 25, 64, and 160 were used throughout, depending primarily upon the light conditions prevailing at the time. Each film has its strengths and its limitations; on the whole, I have no strong preferences. All processing was done by Kodak.

In making exposures, I rely upon a built-in light meter, supplemented by a certain amount of intuition. I neither record nor remember the settings I use. The picture notes that follow are therefore more general than specific regarding these matters, leaning toward the natural history of the subjects rather than the photographic technicalities.

The actual places that comprise the setting for this book are incidental, for the same kinds of pictures could have been taken in any grassland, even in a city weedlot. Because of the circumstance of residence, the majority of the photographs were taken in Marin County, which lies to the north across the Golden Gate Bridge from San Francisco. Some were taken north of Marin in Sonoma County, a few along the edge of the inner coast range and the Central Valley, and many in the Mother Lode country of the Sierra Nevada foothills. Localities are given in the notes for each picture.

Photographic details

PAGES 2-3: Mule deer does at sunrise, near Nicasio, Marin County. 500mm Vivitar; High Speed Ektachrome.

PAGES 4-5: Soldier beetle, one-half inch long, against rising sun in San Geronimo Valley, Marin County. Micro-Nikkor; Kodachrome X.

PAGES 6-7: Tiny ground webs at sunrise, near Ione, Amador County. Micro-Nikkor with M-ring; Kodachrome X.

PAGE 17: Valley oaks, *Quercus lobata*, near El Verano, Sonoma County. Micro-Nikkor; Kodachrome.

PAGES 18 AND 19: Dew on grass and on the delicate meshwork of a dome-web spider, Bear Valley meadow, Point Reyes National Seashore, Marin County. This particular species of spider builds its dome web so low in the grass that it is nearly impossible to get a camera beneath one without disturbing the spider at the same time. I worked all morning, soaked to the skin, in order to get a photograph looking skyward before the spider dropped to the ground. Micro-Nikkor with M-ring; Kodachrome X.

PAGES 20-21: Baby blue-eyes, *Nemophila menziesii*, a pale form from the San Geronimo Valley, Marin County. Micro-Nikkor; Kodachrome X.

PAGES 22 AND 23: California poppies, *Eschscholtzia californica*, from Stinson Beach, and buttercups, *Ranunculus*, from the San Geronimo

Valley, both Marin County. 135mm Soligor with extension tubes; Kodachrome X.

PAGES 24 AND 25: Strecker's day-sphinx, *Pogocolon juanita* (or a near relative) newly emerged, San Geronimo Valley, Marin County. Immature katydid on a flower of *Orthocarpus faucibarbatus* (very similar to Johnny Tuck, *O. erianthus* mentioned in the text), near El Verano, Sonoma County. Both, Micro-Nikkor; Kodachrome X.

PAGE 28: Dry grasses and rocks at sunset, "hill prairie" country on the lower slopes of the inner coast range, west of Los Banos, Merced County. Micro-Nikkor; Kodachrome.

PAGE 29: Dry-grass seed stalks against the rising sun, San Geronimo Valley, Marin County. Micro-Nikkor; Ektachrome.

PAGES 32 AND 33: Seed capsule of the yellow mariposa lily; Micro-Nikkor. Ripgut and other dry grasses against reflection of sunset light on the river; 135mm Soligor. Both taken at Middle Bar, Mokelumne River, Amador County, with Ektachrome.

PAGE 36: Tarweeds, *Hemizonia,* San Geronimo Valley, Marin County. 135mm Soligor; Ektachrome.

PAGE 37. Black blister beetle, in the family Meloidae, feeding on tarweed blossoms, Tiburon Hills, Marin County. Micro-Nikkor; Kodachrome.

PAGE 40: Monarch, *Danaus plexippus,* caught in the web of the banded garden spider, *Argiope trifasciata.* Perhaps because of its large size and violent activity in the web, the spider ignored this victim. After watching for some time, I extricated the butterfly and it flew off, tired but unharmed; Audubon Canyon Ranch, Marin County. 135mm Soligor with a small extension ring; High Speed Ektachrome.

PAGES 42-43: Dew-covered thistle plume stuck to the bud of a tarweed, San Geronimo Valley, Marin County. Micro-Nikkor; Kodachrome X.

PAGES 44 AND 45: Dew-covered dry grasses, San Geronimo Valley, Marin County. Grass-stem shadows behind a fallen wild grape leaf, near Mark West Springs, Sonoma County. Both, Micro-Nikkor; Kodachrome.

PAGES 46 AND 47: Sunset after an autumn rainstorm (LEFT), with a 135mm Soligor; termites emerging (TOP) and new shoots of grass (BOTTOM), with a Micro-Nikkor; all from the San Geronimo Valley, Marin County. Termites with Ektachrome, others with Kodachrome X. Drops of water at the tips of grass blades are not formed of dew. They represent excess water taken in by the roots during the night and extruded from special pores in a process called gutation.

PAGES 48-49: Valley oak, *Quercus lobata,* in a winter morning ground fog, near El Verano, Sonoma County. Micro-Nikkor; Kodachrome X.

PAGES 50 AND 51: Frost on a wild mustard leaf, and ice extruded from pores in the soil as "frost heave" pushing the surface pebbles an inch

or more above the ground; both taken in the San Geronimo Valley, Marin County. Micro-Nikkor; Kodachrome X.

PAGE 52: Spider web covered with frost at sunrise; High Speed Ektachrome. The night was so cold that the spider stopped spinning once the foundation lines were established.

PAGE 53, TOP: Frosted mushrooms at dawn, High Speed Ektachrome pushed to ASA 400. BOTTOM: Buttercup fringed with frost crystals, Kodachrome X. Both taken in the San Geronimo Valley, Marin County, with Micro-Nikkor.

PAGE 56: Orb web, San Geronimo Valley, Marin County. 135mm Soligor with a small extension ring; Ektachrome.

PAGES 58 AND 59: Young katydid and adult crane fly on brome-grasses going to seed, Middle Bar, Mokelumne River, Amador County. Both, Micro-Nikkor; Kodachrome X.

PAGES 62 AND 63: Stages in the emergence of the adult scorpion fly, *Bittacus.* Newly emerged (TOP) and drying its wings (BOTTOM), both from Middle Bar, Mokelumne River, and feeding on a crane fly (RIGHT), near Ione, Amador County. All with Micro-Nikkor; Kodachrome X.

PAGE 65: Harvestman cleaning a hind leg as it retreats to the grass-roots shadows for the day (TOP), and a face view of a tarantula (BOTTOM). Both taken at dawn at Middle Bar, Mokelumne River, Amador County. Micro-Nikkor; Kodachrome X.

PAGE 66: Superstructure of a dome web, Bear Valley, Point Reyes National Seashore, Marin County, at dawn. Micro-Nikkor and M-ring; Kodachrome X.

PAGE 67, TOP: Bowl and doily spider hanging between the bowl and the doily, Middle Bar, Mokelumne River, Amador County, at dawn. Micro-Nikkor and M-ring; High Speed Ektachrome. BOTTOM: Complete snare of the triangle spider (note the spider along the line from web to bush), San Geronimo Valley, Marin County, at sunrise. Micro-Nikkor; Kodachrome X.

PAGES 68-69: One of the many species of orb weavers that inhabit the grasslands, photographed before sunrise at Bear Valley meadow, Point Reyes National Seashore, Marin County. Micro-Nikkor; Kodachrome X.

PAGES 70 AND 71: Golden garden spider, or orange argiope, *Argiope aurantia,* among water parsley in a meadow bog (PAGE 70), and feeding on a wrapped-up honey bee or yellow jacket in a coyote bush (PAGE 71). Note the two small milichiid flies sharing a meal with the spider. Both taken in the San Geronimo Valley, Marin County. 135mm Soligor and an extension ring; Ektachrome.

PAGES 74 AND 75: Courting males (PAGE 74), and a single male approaching the larger female golden garden spider in a coyote bush (PAGE 75), San Geronimo Valley, Marin County. Both, 135mm Soligor plus extension rings; Ektachrome.

PAGES 78 AND 79: Egg cocoon of the golden garden spider in the fall (PAGE 78), and newly hatched spiders in the spring (PAGE 79). Cocoon in the San Geronimo Valley, with Ektachrome; young in Bear Valley meadow, Point Reyes National Seashore, Marin County, with Kodachrome. Both with Micro-Nikkor.

PAGE 82, TOP: Healthy dung fly, Scatophaga, perched on a grass blade in wait for some small prey to wander near. San Geronimo Valley, Marin County. BOTTOM: A carcass of the same species of fly, attached to a fiddle-neck stalk, having been killed by the parasitic fungus, Entomophthora. Pink spore cases of the fungus can be seen protruding as a mass between the segments of the fly's abdomen. Middle Bar, Mokelumne River, Amador County. Both, Micro-Nikkor; Kodachrome X.

PAGE 86, TOP: Snipe fly, family Rhagionidae, waking up at sunrise; San Geronimo Valley, Marin County. BOTTOM: Pacific tree frog, Hyla ragilla, in a lupine leaf near the upper end of Tomales Bay, Marin County. Both, Micro-Nikkor; Kodachrome X.

PAGE 87, TOP: Sparrow hawk portrait, San Geronimo Valley, Marin County; 135mm Soligor. BOTTOM: Garter snake, Middle Bar, Mokelumne River, Amador County; Micro-Nikkor. Both with Kodachrome X.

PAGE 89: Mule deer doe, looking through a coast live oak, near Nicasio, Marin County. 500mm Vivitar; High Speed Ektachrome.

PAGE 90, TOP: Mountain lion, or puma, named Socrates, photographed at the Boyd Science Museum, San Rafael, Marin County. This is the only picture in the book taken outside the natural environment. If man had left these magnificent creatures alone, they would still be common enough in the grasslands to photograph readily in the wild. 135mm Soligor. BOTTOM: Female mosquito just finished taking a meal of blood (mine), upper end of Tomales Bay, Marin County. Micro-Nikkor with extension tubes. Both, Kodachrome X.

PAGE 91, TOP: Yellow jacket chewing meat from the carcass of a deer, San Geronimo Valley, Marin County; Micro-Nikkor. BOTTOM: Yellow jacket returning to the entrance of its nest in the ground; 135mm Soligor and a small extension ring. Both, Kodachrome X.

PAGE 92: Carcass of a mule deer fawn, killed by a car, in the San Geronimo Valley, Marin County. Micro-Nikkor; Kodachrome X.

PAGE 93: Mushroom at the edge of a bishop pine forest, Point Reyes National Seashore, Marin County. Micro-Nikkor; Kodachrome X.

PAGES 94 AND 95: Frozen rodent breath, formed from the condensation of water vapor in warm breath rising from the burrows of meadow mice during a night of 15° F; San Geronimo Valley, Marin County. Micro-Nikkor; Kodachrome X.

PAGE 96: Flower-visiting nitidulidae, or sap beetle of the genus Amartus, feeding on pollen, and the larvae of a small moth eating petals, both within a California poppy blossom in the San Geronimo Valley, Marin County. Micro-Nikkor, probably with a small tube; Kodachrome.

PAGES 98 AND 99: Seed head of a dandelion (PAGE 98), Lake Lagunitas, Marin County. Micro-Nikkor; Kodachrome. Sparrow eating the seeds of an introduced species of thistle (PAGE 99), Audubon Canyon Ranch, Marin County. 135mm Soligor; High Speed Ektachrome.

PAGES 102 AND 103: Thistle seeds fallen neatly upon the ground from a dried flower head; Micro-Nikkor, Ektachrome. PAGE 103: A single seed against the sun, just ready to take off; Micro-Nikkor, with a neutral density filter to reduce exposure factor; Kodachrome. Both, the San Geronimo Valley, Marin County.

PAGES 106-107: Thistle plume photographed almost directly into the sun in order to catch the refraction of light along the hairs; San Geronimo Valley, Marin County. Micro-Nikkor; Ektachrome.

PAGES 110 AND 111: Spiderling balloon and thistle plume discarded among the dry grasses (LEFT); Ektachrome. Thistle seed sprouting after the first rain (TOP), and the web of a young spider settled down after ballooning in the fall with poison oak and snow berry in the background (BOTTOM); both with Kodachrome X. San Geronimo Valley, Marin County; Micro-Nikkor.

PAGE 113: Wyethia, or mule-ears, a genus of the sunflower family with many species in the West. Flowers taken in the spring (TOP), leaf close-up in the summer (BOTTOM), San Geronimo Valley, Marin County. Micro-Nikkor; Kodachrome.

PAGES 114 AND 115: Wyethia, San Geronimo Valley, Marin County. Autumn leaf with shadow of wild oats (LEFT), and winter leaf skeleton edged with frost at sunrise (RIGHT). Micro-Nikkor, with M-ring for frost detail; Kodachrome X.

PAGE 116: Wyethia leaf skeleton and sprouting grass, near Ione, Amador County. Micro-Nikkor; Kodachrome.

PAGES 118 AND 119: Oregon oak, Quercus garryana, in a winter fog near Mark West Springs, Sonoma County. Nikkor-S 50mm; Kodachrome.

PAGE 120: Acorn of coast live oak, Quercus agrifolia, sprouting from the bark of a Douglas-fir, storage tree for a colony of acorn woodpeckers in Bear Valley meadow, Point Reyes National Seashore, Marin County. 135mm Soligor; Kodachrome II.

PAGE 122: Fishnet lichen in a coast live oak, Quercus agrifolia, photographed by Maggie Cavagnaro near Mark West Springs, Sonoma County; 135mm Soligor, Ektachrome.

PAGE 123, TOP: Caterpillar chewing on the leaves of blue oak, Quercus douglasii; Micro-Nikkor. BOTTOM: "Oak apples," stem galls formed on the twigs of valley oak, Quercus lobata, by cynipid wasps; 135mm Soligor. Both at the edge of the Sacramento Valley west of Winters, Yolo County; Kodachrome.

PAGE 126: Madrone bark peeling at the end of summer, near Gualala, Mendocino County. 135mm Soligor; Ektachrome.

PAGE 127: Moss-covered limbs of a valley oak during the winter, San Geronimo Valley, Marin County. 135mm Soligor; Kodachrome.

PAGE 130: Great horned owl, photographed by Maggie Cavagnaro at Sunol Valley Regional Park, Alameda County. 135mm Soligor; Ektachrome.

PAGE 131: Coast live oak, *Quercus agrifolia*, against the setting sun, near Nicasio, Marin County. 500m Vivitar; Ektachrome.

PAGE 134: Dewdrops on the leaflets of lupine, photographed almost directly into the sun, San Geronimo Valley, Marin County. Micro-Nikkor with M-ring; Kodachrome.

PAGES 138 AND 139: Raindrops on grass blades, San Geronimo Valley, Marin County. PAGE 139: Bare branches of valley oak, winter sunrise near El Verano, Sonoma County. Both, Micro-Nikkor; Kodachrome.

PAGES 142 AND 143: Distant and close views of the same seed stalk of rye grass draped with a dew-covered sheet web, San Geronimo Valley, Marin County. Micro-Nikkor, with M-ring for close-up; Ektachrome.

PAGES 146-147: An orb web covered with dew, back-lit at sunrise with a shadow in the background; Bear Valley, Point Reyes National Seashore, Marin County. Micro-Nikkor; Kodachrome.

PAGES 150 AND 151: Spider web spectra, caused by light refracting from the web structure. The clover bud is about one-half inch across, the spider about one-quarter inch. Photographing these rainbows is difficult at best. All dew must have evaporated from the web, but the strands seem to be most colorful before the heat of the day has dried them out entirely. The lens must be pointed almost directly at the sun, yet carefully shaded, in order to be aligned properly to see the refracted light. The web images that are spread by being out of focus exhibit the best rainbows. Deer Park, Fairfax, Marin County. Micro-Nikkor; Kodachrome.

PAGE 153: Hercules, the jackrabbit we raised from infancy and then released in Samuel P. Taylor State Park, Marin County. Photographed in the evening. 135mm Soligor; High Speed Ektachrome.

PAGES 154 AND 155: Young fiddlehead of bracken fern, and the shell of the common European garden snail, now widely established in California. Both taken in the San Geronimo Valley, Marin County. Micro-Nikkor with M-ring for snail; Kodachrome.

PAGES 156 AND 157: Filaree seeds, *Erodium*, in three stages of development: Still green on the plant (LEFT), twisting off the central stem, as seen from above (TOP), and planting themselves in the ground (BOTTOM). All taken at Middle Bar, Mokelumne River, Amador County. Micro-Nikkor; Ektachrome and, for planted seeds, Kodachrome.

PAGES 158 AND 159: Anise swallowtail, feeding on fiddleneck flowers (LEFT), and detail of the underside of the hind wing; Middle Bar, Mokelumne River, Amador County. Micro-Nikkor; Kodachrome and, for the close-up, Ektachrome.

PAGE 160: Body contour feather of unknown origin, shed in the dry grass during summer molt, photographed at sunset, Middle Bar, Mokelumne River, Amador County. Micro-Nikkor; Ektachrome.

PAGE 162: A tiny gnat, about one-eighth inch in length, photographed at sunrise against the sun, near Ione, Amador County. Micro-Nikkor and extension tubes; Kodachrome X.

PAGE 163: Slug returning to the grass-roots darkness as the sun rises over Bear Valley meadow, Point Reyes National Seashore, Marin County. Micro-Nikkor; Kodachrome X.

PAGE 166: Blossom of turpentine weed, *Trichostema lanceolatum*, San Geronimo Valley, Marin County. Micro-Nikkor; Ektachrome.

PAGE 167: Indian soaproot, *Chlorogalum pomeridianum*, a summer-blooming lily, the flowers of which are pollinated at night by moths. The bumblebee had been out gathering pollen by day and, failing to reach the nest as evening approached, held firmly to a twig to sleep for the night. Audubon Canyon Ranch, Marin County. Micro-Nikkor; Kodachrome X.

PAGE 170: "Hill prairie" country where the San Joaquin Valley meets the inner coast range, west of Los Banos, Merced County. In an area of very low rainfall, this grassland was already brown in early May. 135mm Soligor; Kodachrome.

PAGE 171: Wild oats, *Avena fatua*, one of the many annual European grasses that now dominate the grasslands of California, with sun's reflection off the river coming through an oak in the background. Middle Bar, Mokelumne River, Amador County. 135mm Soligor and small extension ring; Ektachrome.

PAGE 174: A portrait of Maggie watching a small orb weaver on its web; San Geronimo Valley, Marin County. 135mm Soligor; Kodachrome.

PAGES 178-179: Blue oak, *Quercus douglasii*, at sunrise, edge of the Sacramento Valley, just west of Winters, Yolo County. Micro-Nikkor; Kodachrome.

PAGES 182-183: Spider web reflected in a drop of water extruded by the grass blade by gutation. This was a lucky discovery that offers its own interesting lessons in nature photography. The picture was taken in an unusual orb web draped horizontally over the grasses rather than suspended vertically. Because of this arrangement, Jonathan Braun, our companion on a morning walk at Middle Bar in Amador County, spotted the web and called me over. I photographed the web for ten or fifteen minutes, lying on my belly in the wet grass. I took half a roll of pictures from all angles, starting with the whole web and moving in closer and closer. Two shots before I ran out of film, I saw the reflection in the drop, though my nose had been a few inches from it all the time. One shot was underexposed; this picture was the last one on the roll. Micro-Nikkor with M-ring; Kodachrome; perseverance and lots of luck.

INDEX

Body type, 11-point Bembo; display faces, Arrighi and Centaur. Composition by Graphic Arts Center of Portland, Oregon, and Mackenzie & Harris, Inc., San Francisco, Calif. Color separations, printing, and binding by Kingsport Press, Inc., Kingsport, Tennessee.

Design by David Cavagnaro and Arthur Andersen.